Small-scale manufacture
of stabilised soil blocks

TECHNOLOGY SERIES Technical Memorandum No. 12

Small-scale manufacture of stabilised soil blocks

Prepared under the joint auspices of the International Labour Office
and the United Nations Industrial Development Organisation

International Labour Office   Geneva

Copyright © International Labour Organisation 1987

Publications of the International Labour Office enjoy copyright under Protocol 2 of the Universal Copyright Convention. Nevertheless, short excerpts from them may be reproduced without authorisation, on condition that the source is indicated. For rights of reproduction or translation, application should be made to the Publications Branch (Rights and Permissions), International Labour Office, CH-1211 Geneva 22, Switzerland. The International Labour Office welcomes such applications.

ISBN 92-2-105838-7

ISSN 0252-2004

First published 1987

The designations employed in ILO publications, which are in conformity with United Nations practice, and the presentation of material therein do not imply the expression of any opinion whatsoever on the part of the International Labour Office concerning the legal status of any country, area or territory or of its authorities, or concerning the delimitation of its frontiers.
The responsibility for opinions expressed in signed articles, studies and other contributions rests solely with their authors, and publication does not constitute an endorsement by the International Labour Office of the opinions expressed in them.
Reference to names of firms and commercial products and processes does not imply their endorsement by the International Labour Office, and any failure to mention a particular firm, commercial product or process is not a sign of disapproval.

ILO publications can be obtained through major booksellers or ILO local offices in many countries, or direct from ILO Publications, International Labour Office, CH-1211 Geneva 22, Switzerland. A catalogue or list of new publications will be sent free of charge from the above address.

Printed by the International Labour Office, Geneva, Switzerland

PREFACE

This technical memorandum on small-scale production of stabilised soil blocks is the ninth in a series of memoranda currently being prepared by the ILO and UNIDO.[1] It is the second of three memoranda on building materials for low-cost housing.[2]

This technical memorandum is of particular importance to developing countries in view of the current severe shortage of shelter for large sections of the population in these countries. Yet, after food and clothing, adequate shelter is one of the most important basic needs. It is estimated that one-fourth of the world's population does not have adequate housing. An average of 50 per cent of urban populations live in slums. In some developing countries, urban slums constitute up to 80 per cent of urban settlements. The housing situation in developing countries will further deteriorate unless substantial resources are allocated to this sector by governments and international aid. This explains the decision of the United Nations General Assembly formally to proclaim 1987 as the International Year of Shelter for the Homeless (IYSH), with a view to securing renewed political commitment and effective action within and among the international community. The International Labour Office will contribute, in the future, to the achievement of the above objectives, especially since the implementation of appropriate housing policies will also generate a great number of much needed employment opportunities. It is hoped that the preparation and dissemination of this memorandum will be helpful in formulating such policies.

Developing countries wishing to expand substantially the housing stock for low-income groups will have to identify the least costly solutions, in terms of unit housing cost and the foreign exchange content of such cost. Furthermore, these solutions should allow, whenever possible, the direct involvement of potential home owners who may wish to contribute their labour in, for example, self-help housing schemes. The use of soil as an alternative building material for a wide range of housing types should be part of these solutions, and should be promoted by housing authorities for the following reasons.

Firstly, soil is already being used as a main building material by a very large number of developing countries but is often considered as a second-best or poor-man's solution. Thus, whenever financially possible, there is a tendency to switch to other building materials which are considered more "modern" (e.g. concrete) than soil. It is therefore important to reverse this trend by demonstrating that properly processed soil is as good as, or even better than, these modern materials.

Secondly, houses built with blocks of stabilised soil are often less expensive than those built with other materials, such as concrete blocks or wood. Thus, the use of soil should facilitate home ownership and minimise government subsidies for low-cost housing projects.

Thirdly, the use of soil requires substantially fewer imported inputs than many other building materials, and should therefore contribute to an improvement in the balance of payments situation of developing countries.

Fourthly, the building of a housing unit with stabilised soil will often generate more direct and indirect employment than if the same housing unit was built with other materials, such as concrete or fired bricks.

Finally, houses made of stabilised soil often offer a more pleasant environment (e.g. in terms of protection against outside heat or cold) than houses made of the so-called "modern" materials.

From many points of view - technical, cultural, environmental, financial - soil could be given preference as a building material. In order to expand the use of this material, housing authorities will need to implement three groups of measures. The first group relates to the improvement of housing design and construction processes. It has now been proved that soil can be a sound building material if properly used. A large number of experiments have been successfully conducted throughout the world, the techniques and tools have been improved and technical solutions have been found for the three main problems which militated against the use of soil as a building material: the deterioration of earth walls by rain; low resistance to earthquakes; and the difficulty of building floor slabs. Furthermore, the maintenance of earth buildings may be considerably reduced and their lifespan increased if appropriate designs are used and raw materials adequately processed. A large number of construction projects completed under a wide range of climatic conditions in both developing and developed countries demonstrate that there are currently no unsolvable problems in the use of earth as a building material. Housing authorities therefore need to promote research in this field, disseminate technological information on earth building techniques, and provide training facilities - at all levels - for the proper processing and use of soil for building.

The second group of measures relates to the government policies required to induce individuals and contractors to adopt soil-based materials in housing projects. Housing authorities would need to advise the central Government to formulate and implement fiscal and monetary measures in favour of the adoption of earth as an alternative building material. For example, higher duties could be applied on imported materials and higher housing subsidies could be granted for earth buildings. Preference could be given to contractors bidding for government-financed projects (e.g. construction of schools) whenever they offer to use earth as the main building material.

The third group of measures relates to the dissemination of information on the utilisation of earth as a building material. Such information should dissipate doubts on the technical and economic efficiency of this material and improve the image of earth buildings among those who may feel that the use of earth for building purposes is a second-best solution for countries which may not be able to afford the so-called "modern" materials. It is hoped that the information contained in this technical memorandum will help achieve these goals.

The ILO is not the only institution promoting the use of soil for buildings. Currently, a large number of centres in both developing and developed countries are vigorously promoting this material for all types of building: low and middle-income housing, luxury houses, office buildings, religious buildings, and so on. These centres are located both in the North and in the South, on all continents and under a wide range of climates (see Appendix II). The proliferation of such centres is indicative of the renewed interest in earth as an alternative building material. It is interesting to note that a few days before this memorandum was being sent for print, the use of earth as a building material was the main topic of a popular television programme in France.[3]

As in the case of the other technical memoranda, the main objective of this memorandum is to provide small-scale producers in developing countries with detailed technical information on small-scale technologies which have been successfully applied in a number of countries, but are not well known in others. A secondary objective is to assist public planners in identifying and promoting technologies consonant with national socio-economic objectives, such as employment generation, foreign exchange savings, rural industrialisation, or the fulfilment of the basic needs of low-income groups.

The information contained in this memorandum is sufficiently detailed for small-scale producers to identify and apply the technologies described in the text without the need for further information. Thus, detailed drawings of equipment which may be manufactured locally are provided and a list of equipment suppliers from both developing and developed countries is annexed in order to help producers choose the equipment which must be imported. In the few instances where the available information is not sufficient, the reader may obtain additional technical details from publications listed in the bibliography.

Technical memoranda are not intended as training manuals. It is assumed that the potential users of the technologies described therein are trained practitioners, and that the memoranda are only supposed to provide them with information on alternative technological choices.

This technical memorandum contains eights chapters, five of which deal with the various sub-processes needed for the manufacture of stabilised soil blocks, including quarrying and testing of raw materials; pre-processing of the latter (grinding, sieveing, proportioning, and mixing); block forming methods including a detailed description of alternative block forming machines; curing and testing of blocks; and the use of mortars and renderings in wall construction. The last chapter (Chapter VIII) is mostly intended for public planners and project evaluators from industrial development agencies who wish to obtain information on the various socio-economic effects of the production and use of alternative building materials.

The memorandum also contains four appendices which could be of interest to the reader. Appendix I provides a glossary of technical terms, and should therefore be of assistance to non-specialists. Appendix II provides a list of institutions from which additional information on earth building techniques may be obtained. Appendix III provides a list of equipment suppliers and manufacturers from both developing and developed countries. It may be noted that this list is far from being exhaustive and that it does not imply a special endorsement of these suppliers and manufacturers by the ILO or UNIDO. The names listed are only provided for illustrative purposes and readers are urged to obtain additional information from as many sources as possible. Appendix IV provides a bibliography on the subject, which may be useful in learning more about the techniques described in the main body of the text.

A questionnaire is attached at the end of the memorandum for those who may wish to send to the ILO or UNIDO their comments and observations on the content and usefulness of this publication. These will be taken into consideration in the preparation of future technical memoranda.

The memorandum was prepared by R.G. Smith and D.J.T. Webb, staff members of the Building Research Establishment (United Kingdom) in collaboration with M. Allal, staff member in charge of the preparation of a series of technical memoranda within the Technology and Employment Branch of the International Labour Organisation. The preparation of this memorandum also benefited from very useful information and suggestions provided by a large number of individuals and institutions. The ILO, UNIDO and the authors acknowledge their generous assistance.

A.S. Bhalla,
Chief,
Technology and Employment Branch.

---

[1] Three other memoranda have been published jointly with FAO and UNEP.
[2] One technical memorandum on small-scale brickmaking (Technical Memorandum No. 6) has already been published. Another memorandum on the small-scale production of windows and doors for low-cost housing will be available in 1987.
[3] This programme, entitled "Ambitions", went on air on 3 December 1986.

CONTENTS

Page

PREFACE.. v

ACKNOWLEDGEMENTS... xv

CHAPTER I.   INTRODUCTION.. 1

    I.   PURPOSE AND OBJECTIVES OF THE MEMORANDUM..................... 1

    II. NEED TO IMPROVE TECHNIQUES FOR PRODUCTION OF
        STABILISED SOIL BLOCKS....................................... 2

    III. COMPARISON BETWEEN STABILISED SOIL BLOCKS AND
        OTHER BUILDING MATERIALS..................................... 3

        III.1 Compressive strength................................... 4
        III.2 Moisture movement...................................... 4
        III.3 Density and thermal properties......................... 6
        III.4 Durability, maintenance and appearance................. 6

    IV. SCALES OF PRODUCTION COVERED BY THE MEMORANDUM............... 8
    V.   CONTENT OF THE MEMORANDUM.................................... 10
    VI. TARGET AUDIENCE.. 11

CHAPTER II. RAW MATERIALS, TESTING AND STABILISERS...................... 13

    I.   RAW MATERIALS.. 13
    II. QUARRYING THE RAW MATERIAL................................... 16
    III. SOIL TESTING PROCEDURES...................................... 23
        III.1 Preliminary on-site tests.............................. 23
        III.2 Further soil testing procedures........................ 26
        III.3 Laboratory testing methods............................. 31
    IV. SOIL STABILISERS... 37
        IV.1 Principles of soil stabilisation...................... 37
        IV.2 Soil stabilisation methods............................ 38

CHAPTER III. PRE-PROCESSING OF RAW MATERIALS	45
I. THE NEED FOR PRE-PROCESSING	45
II. GRINDING	46
II.1 Simple hand tools	46
II.2 Pendulum crusher	46
II.3 Other hand-powered methods	50
III. SIEVING	52
IV. PROPORTIONING	52
V. MIXING	54
VI. PRODUCTIVITY OF LABOUR AND EQUIPMENT	55
VII. QUANTITY OF MATERIALS REQUIRED	55
CHAPTER IV. FORMING	57
I. BUILDING STANDARDS AND BLOCKS	57
II. COMPRESSIVE STRENGTH: DENSITY AND MOULDING PRESSURE RELATIONSHIPS	58
III. BLOCK-FORMING METHODS	61
IV. SOIL TESTING PRIOR TO PRODUCTION	62
V. BLOCK SIZES	64
VI. PROPOSED TECHNICAL STANDARDS FOR COMPRESSED SOIL BLOCKS	65
VII. SOIL BLOCK MAKING MACHINES	69
VII.1 The CINVA-Ram press	70
VII.2 The CETA-Ram press	73
VII.3 Landcrete press/Presse Terstaram	75
VII.4 Tek-Block press	75
VII.5 Winget block making machine	78
VII.6 Ellson Blockmaster stabilised soil block press	80
VII.7 Consolid AG	82
VII.8 Supertor block making machine	84
VII.9 Maquina block making machine	84
VII.10 Brepak block making machine	84
VII.11 Zora hydraulic block press	85
VII.12 Latorex system	88
VII.13 Astram block making machine	88

	VII.14 Tecmor equipment	90
	VII.15 Meili 60 manual soil block press	91
	VII.16 Terrablock Duplex machine	91
VIII.	WORLD SURVEY OF SOIL BLOCK MAKING EQUIPMENT	92

CHAPTER V. CURING AND TESTING... 99

    I. INTRODUCTION... 99
    II. THE NEED FOR CURING AND TESTING......................... 99
    III. METHODS OF CURING...................................... 101
    IV. TESTING STABILISED SOIL BUILDING BLOCKS................. 103
        IV.1 Site testing procedures............................ 103
        IV.2 Laboratory testing methods......................... 109
        IV.3 Durability tests................................... 113
        IV.4 Long term exposure tests........................... 115
        IV.5 Selection of an exposure site...................... 117

CHAPTER VI. MORTARS AND RENDERINGS.. 119

    I. NEED FOR MORTARS AND RENDERINGS......................... 119
    II. MORTAR TYPES... 120
    III. TYPES OF RENDERING.................................... 123
    IV. MIXING AND USE... 124

CHAPTER VII. COSTING.. 127

    I. VARIATIONS IN COSTS..................................... 127
    II. METHODOLOGICAL FRAMEWORK............................... 128
    III. APPLICATION OF THE METHODOLOGICAL FRAMEWORK........... 131

CHAPTER VIII. SOCIO-ECONOMIC CONSIDERATIONS............................... 139

    I. INTRODUCTION.. 139
    II. ACCEPTANCE AND APPLICATION............................. 139
    III. EMPLOYMENT GENERATION................................. 141
    IV. INVESTMENT COSTS AND FOREIGN EXCHANGE SAVINGS.......... 142
    V. PRODUCTION COST OF STABILISED SOIL BLOCK
       AND BUILDING COSTS...................................... 143
    VI. CONCLUDING REMARKS.................................... 146

APPENDICES

APPENDIX I.	GLOSSARY OF TECHNICAL TERMS	151
APPENDIX II.	INFORMATION SOURCES ON STABILISED SOIL BLOCKS	159
APPENDIX III.	LIST OF EQUIPMENT SUPPLIERS AND MANUFACTURERS	169
APPENDIX IV.	BIBLIOGRAPHY	175

QUESTIONNAIRE... 179

ACKNOWLEDGEMENTS

The publication of this technical memorandum has been made possible by a grant from the Overseas Development Administration of the United Kingdom, through the Intermediate Technology Development Group (ITDG, London and Rugby). The International Labour Office and the United Nations Industrial Development Organisation acknowledge this generous support. The photographs have been prepared specifically for this Memorandum by the Building Research Establishment.

CHAPTER I

INTRODUCTION

I. **PURPOSE AND OBJECTIVES OF THE MEMORANDUM**

Housing is one of the most important basic needs of low-income groups in developing countries. However, it is a most difficult need to satisfy, since land and building costs are often outside the means of both the rural and urban poor. Thus, many governments have launched various schemes with a view to facilitating some form of housing ownership by low-income groups, including self-help housing schemes, housing subsidies, provision of credit, low interest rates and so on.

In view of the limited means at the disposal of governments and potential home owners, it is important to seek ways to lower construction costs of low-income housing while minimising repair and maintenance costs. This can be achieved partly through the production and use of cheap yet durable building materials, since these usually represent a very large proportion of total low-income housing costs in developing countries. Furthermore, it would be useful if the production of these building materials could contribute to the fulfilment of important development objectives of these countries, such as the generation of productive employment, rural industrialisation and a decreased dependence on essential imports.

A number of traditional building materials exist which have proved to be the most suitable for a wide variety of buildings and which have a great potential for increased use in the future. These building materials, which are made from locally available raw materials, can be produced close to or at the construction site, with little equipment (which may be produced by local rural or urban workshops), and are often more appropriate to the environment than alternative "modern" materials such as cement or plastic-based materials. One such building material is the stabilised soil block, a modified form of one of the oldest materials used in housing construction.

The purpose of this technical memorandum is to provide detailed technical and economic information on small-scale production of stabilised soil blocks with a view to assisting those who produce or plan to manufacture these commercially - in self-help housing schemes or housing co-operatives - to improve their production techniques and the quality of the output. It is also hoped that the information contained in this memorandum will induce governments to promote greater use of this material for middle and high-income housing, as is starting to be the case in some industrialised countries (e.g. France, where the building of adobe housing is gaining prominence). Such a step will have a very significant effect on a country's balance of payments, since it will reduce the import of expensive building materials and of the inputs and equipment needed for the local production of similar materials (e.g. cement, energy, turn-key factories for the production of bricks and cement-based materials).

II. NEED TO IMPROVE TECHNIQUES FOR PRODUCTION OF STABILISED SOIL BLOCKS

Soil has been used in the construction of human shelters for thousands of years. In countries characterised by relatively humid and rainy weather, soil is not, in itself, a particularly durable building material. Thus, much effort is usually expended in protecting and repairing soil structures. If soil is to be used more frequently as a building material, its performance must be improved in order to make it as attractive, or more attractive than, alternative materials. This may be achieved in two main ways. First, soil can be made more resistant to water if it is mixed, for example, with bituminous compounds. Second, the nature of soil can be modified in order to improve its durability if it is mixed with lime or other additives.

Soil may either be built into a wall in situ or moulded into building blocks. In the first method, walls may be built in three different ways. In "cob" construction, walls are built by placing handfuls of moist soil, layer by layer. Alternatively, a strengthening framework of wooden sticks is built and filled with moist soil (wattle and daub), or soil is rammed with a heavy weight into the space between a pre-erected formwork, as in pisé de terre.

The second method consists in fashioning the moist soil into building blocks, which are then used in wall construction with mortar.

In the _in situ_ method, the drying shrinkage takes place within the wall. This is not the case for stabilised soil blocks, which are allowed to dry and shrink before usage, thus minimising the risk of cracks in the finished structure. Blocks can give an excellent finish to a wall surface.

In many countries, the quality of the stabilised soil blocks used in some housing schemes is far from adequate. Furthermore, the production of such blocks is sometimes wasteful of materials, such as the stabilisers used in the production of these blocks. If there is to be increased use of these blocks in all types of housing (e.g. low-cost housing in rural and urban areas; middle-income housing in urban areas), an improvement in the production technique - aimed at improving quality and reducing production costs - will be required. In order to improve the production technique, the following will need to be carefully considered:
- adoption of optimum proportions of stabilised soil and other inputs, taking into consideration the characteristics of local soil;
- careful mixing of the various components of stabilised soil blocks;
- application of an adequate compaction pressure to the moist soil in order to obtain dense and strong building blocks with well-shaped surfaces and edges; this will require the use of efficient block making machines;
- obtaining a smooth block surface in order to allow the use of blocks without rendering or with a minimum use of rendering materials.

Good quality stabilised soil blocks should improve hygiene (e.g. there will be no cracks on the surface for insects to lodge in), reduce housing maintenance and repair costs and, in general, extend the life of a building.

The following chapters provide technical information which should help established or potential small-scale producers to apply appropriate techniques in the various stages of stabilised soil block making with a view to improving quality and reducing production costs.

III. COMPARISON BETWEEN STABILISED SOIL BLOCKS AND OTHER BUILDING MATERIALS

This section compares the characteristics of stabilised soil blocks with those of other walling materials. This comparison should be useful for housing authorities and builders' associations who must choose among various building materials for specific housing programmes or public buildings. The properties of these materials are summarised in table I.1.

III.1 Compressive strength

The <u>compressive strength</u> of stabilised soil blocks (i.e. the amount of pressure they can withstand without being destroyed) depends upon the nature of the soil, the type of stabiliser used and the pressure applied to form the block. Highest strengths (expressed in MN/m^2)[1] are obtained when the mixing of components and curing (or autoclaving) are carried out properly, and ideal materials are available. In practice, typical wet compressive strengths of stabilised soil blocks may be less than 4 MN/m^2. Such strength is adequate for many building purposes. It compares favourably, for example, with the minimum British Standard[2] requirements of 2.8 MN/m^2 for precast concrete masonry units and load-bearing fired clay blocks, and of 5.2 MN/m^2 for bricks. Where building loads are small (e.g. in the case of single-storey construction), a compressive strength of 1.4 MN/m^2 may be sufficient. This figure is recommended by several building authorities throughout the world.

III.2 Moisture movement

Porous building materials of the type used for wall building may expand slightly when wet and contract again as they dry out. Cracking, spalling and other defects may result in a building if there is excessive movement of the materials.

---

[1] The abbreviation MN/m^2 means "mega newtons per square metre" (i.e. million newtons per square metre). The newton is a unit of force defined as follows: a force which, when acting for one second on a mass of one kilogram gives it a velocity of one metre per second. Compressive strengths are also expressed in pounds per square inch, where one pound per square inch is equal to 6,894.7 newtons per square metre. 1.0 MN/m^2 is equivalent to the following:
1.0 MN/m^2 = 1 N/mm^2 = 1 Mpa = approx. 10 kg f/cm^2 = approx. 145 lb/sq.in.
[2] See British Standards Institution, BS6073, 1981 and BS3921, 1974.

Table I.1
Range of properties of stabilised soil blocks and alternative walling materials

Property[1]	Stabilised soil blocks	Fired clay bricks	Calcium silicate bricks	Dense concrete bricks	Aerated concrete blocks	Light-weight concrete blocks
Wet compressive strength (MN/m^2)	1-40	5-60	10-55	7-50	2-6	2-20
Reversible moisture movement (per cent linear)	0.02-0.2	0-0.02	0.01-0.035	0.02-0.05	0.05-0.10	0.04-0.08
Density (g/cm^3)	1.5-1.9	1.4-2.4	1.6-2.1	1.7-2.2	0.4-0.9	0.6-1.6
Thermal conductivity ($W/m°C$)	0.5-0.7	0.7-1.3	1.1-1.6	1.0-1.7	0.1-0.2	0.15-0.7
Durability under severe natural exposure	Good to very poor	Excellent to very poor	Good to moderate	Good to poor	Good to moderate	Good to poor

[1] See sections III.1 and III.4 for a definition of the properties of materials.

Some soils tend to expand and contract a great deal and are not, therefore, very suitable for earth construction. However, the addition of a stabiliser will reduce such movement. Nevertheless, there may be greater movement in buildings constructed of stabilised soil blocks than in those constructed with alternative materials (see table I.1). Good production, curing and construction methods will minimise such movement. Moisture movement is expressed in terms of linear per cent changes.

It may be noted that such movement becomes especially significant when two materials with different movement characteristics are in close juxtaposition in a building. Differential movement gives rise to stress which may be sufficient to break the bond between the materials, or lead to other damage. For example, cement renderings often become detached from soil blocks which have not been properly stabilised.

III.3 Density and thermal properties

Stabilised soil blocks are generally denser than a number of concrete materials, such as aerated and lightweight concrete blocks, while exhibiting densities similar to those of various types of bricks (e.g. clay, calcium silicate and concrete bricks - see table I.1). While the high density of stabilised soil blocks may be a disadvantage when they have to be transported over long distances, it is of little consequence when blocks are produced at, or close to, the construction site, a fairly common occurrence in earth construction. Furthermore, the high density of stabilised soil blocks has the advantage over lightweight building materials of greater thermal capacity. This characteristic is particularly sought in the tropics where extremes of temperatures are moderated inside buildings made of soil blocks.[1]

III.4 Durability, maintenance and appearance

Soil blocks containing stabilisers show improved weather resistance. Experiments in Ghana, with various proportions of lime have shown marked differences between test walls made of unstabilised and stabilised blocks. Figures I.1 and I.2 illustrate the difference after three years of exposure: unstabilised blocks have been eroded while stabilised blocks were not. An experimental building constructed of bricks made from silty sand and five per cent cement was reported to be in good condition after 23 years of service in the temperate climate of the United Kingdom. Figure I.3 illustrates the excellent condition of part of the walling after 33 years.

Well-made stabilised soil blocks can compare favourably with other walling materials and require little maintenance over a long period of time.

The appearance of the blocks depends upon soil colours, particle size, and the type of process used. Blocks can be made of sufficiently good shape, consistent size, and high quality finish to be built into fair-faced walling. Although the rendering of wall surfaces is traditionally carried out in some countries, it should not be necessary. A white finish, if required to reduce solar gain, may be equally well applied directly to the blocks as to a render coat.

---

[1] See table I.1 which compares the thermal conductivity of soil blocks to that of other materials. This characteristic of building materials is often expressed in watts per metre per degree centigrade.

Figure I.1
Unstabilised soil blocks after 3 year's exposure in Ghana

Figure I.2
Lime-stabilised soil blocks after 3 year's exposure in Ghana

Figure I.3
Wall made of cement-stabilised soil blocks after
33 year's exposure in the United Kingdom

Stabilised soil blocks, in common with other types of blocks and bricks, will require an appropriate amount of steel reinforcement if used in areas of high seismic or high wind risk.

Fire, termites, bacteria and fungi, or ultraviolet radiation should not constitute a hazard for stabilised soil blocks or any other types of blocks. In comparison, organic materials may be at a disadvantage in this respect.

IV. SCALES OF PRODUCTION COVERED BY THE MEMORANDUM

The rate of production of stabilised soil blocks depends, to a large extent, upon the degree of mechanisation of the process. Hand-powered equipment may produce a few hundred units per day, while machine-powered equipment has been developed to produce several thousand units per day. Table I.2 provides the range of outputs for respectively small-scale and large-scale production.

Small-scale production obviates the need for high capital investments and is particularly appropriate in cases where it must satisfy the needs of small, isolated communities.

It will be shown in chapter VIII that small-scale production presents other advantages such as the generation of productive employment, reduction of transport costs, improvement in the balance of payments and the generation of backward linkages (e.g. local production of tools and pieces of equipment).

Table I.2

Scales of production

Scale of production	Number of blocks produced per day	Type of production	Approximate time required to produce enough blocks for a small house
Small	up to 400	Hand-powered	1 week (or more)
Large	400 to 4000	Machine-powered	1 day to 1 week

By contrast, large-scale production may require expensive, relatively sophisticated machines which usually will have to be imported. It will also be shown that large plants generate relatively little employment, may involve greater distribution costs, and require more advanced skill levels for maintenance and repair. Spare parts may have to be imported, and subsequent long delivery times can result in serious production losses. The economies of scale often sought by those who install the larger plants could be realised if production could be maintained and goods sold continuously to a not too distant market.

This technical memorandum focuses primarily on small-scale production for two main reasons. Firstly, technological information of interest to small-scale entrepreneurs is often difficult to obtain, since it is not usually publicised in trade journals or marketed by engineering firms or equipment suppliers. Thus, this memorandum attempts to bridge this information gap. Secondly, detailed information on large-scale plants is outside the scope of a publication of this type. Furthermore, entrepreneurs considering large investments for the establishment of large-scale block making plants will need the services of an engineering firm in view of the risks involved. The memorandum will nevertheless provide a brief description of large-scale plants with a view to providing public planners and housing authorities with a comparison of the socio-economic effects of small and large-scale production.

V. CONTENT OF THE MEMORANDUM

Following the introduction, Chapter II describes the raw materials used in the production of stabilised soil blocks (mainly different types of soil and stabilisers) including their physical and chemical characteristics and the tests used to determine their suitability for block making. Consideration is given to methods of determining the optimum quantities of materials and processing conditions for the production of good-quality blocks.

Chapters III to V describe in detail the various processing stages, including the following:

- breaking down the soil into small particles and mixing it with stabilisers and water;

- forming the blocks (including a description of the various presses available and their effectiveness); estimates of labour requirements are also provided; and

- moist curing and testing of pressed blocks.

Mortars used with the blocks under various conditions are described in Chapter VI. Plastering or rendering of wall surfaces is also discussed in this chapter. Guide-lines to the estimation of unit production costs and the socio-economic aspects to be considered are given in Chapters VII and VIII.

The memorandum concludes with the following appendices:

- glossary of technical terms;
- bibliography;
- list of institutions where information may be obtained; and
- list of equipment suppliers and manufacturers.

VI. TARGET AUDIENCE

This memorandum is intended to provide information to various groups of individuals or institutions concerned with building and construction in developing countries. These include:

- housing authorities concerned with housing construction for low and middle-income groups;
- building research institutes;
- government officers, especially those responsible for housing and public building;
- financial institutions, banks and businessmen;
- small entrepreneurs who may wish either to manufacture blocks at one location for transport to building sites or to move the block making equipment nearer to the source of raw material and the building site;
- builders' co-operatives, such as those formed between would-be house-owners unable to afford the cost of a machine; and
- voluntary organisations, expatriate technical aid workers, extension workers and staff of technical colleges.

CHAPTER II

RAW MATERIALS, TESTING AND STABILISERS

I. RAW MATERIALS

The basic raw material needed to produce stabilised soil building blocks is soil containing a minimum proportion of silt and clay to provide cohesion. Not all soils are suitable for building purposes. The soil characteristics and climatic conditions of the area must be assessed. For example, a dry, semi-desert climate requires different soil blocks from those used in temperate, rainy or monsoon areas.

Soils are variable and complex materials, whose properties can be modified to improve performance in building construction by the addition of various stabilisers.

All soils consist of disintegrated rock, decomposed organic matter and soluble mineral salts. A soil can be graded into fractions according to a system of soil classification widely used in civil engineering. Such classification, based on particle size, is provided in Table II.1:

Table II.1
Soil classification according to particle size[1]

Diameter of particle (mm)	Name of fraction
60 - 20	Coarse gravel
20 - 6.0	Medium gravel
6.0 - 2.0	Fine gravel
2.0 - 0.6	Coarse sand
0.6 - 0.2	Medium sand
0.2 - 0.06	Fine sand
0.06 - 0.02	Coarse silt
0.02 - 0.006	Medium silt
0.006 - 0.002	Fine silt
Less than 0.002	Clay

---

[1] See British Standards Institution, BS1377, 1975.

Soils can also be classified in terms of being heavy or light to work and handle, depending on the texture of the soil. There are seven main types of soil: clay soils, heavy loams, medium loams, sandy loams, sandy soils, chalk and limestone soils, and peat soils. Figure II.1 illustrates the composition of the more common soils with respect to sand and the combined silt and clay content.

It is possible to measure the proportions of silt, sand and clay within a soil, with the help of the triangular diagram represented in figure II.2. This triangular, soil classification chart was originally developed by the Public Roads Administration of the United States. For example, the soil indicated at point X of the chart would be classified as a clay soil with the following constituents: 10 per cent silt; 50 per cent clay; and 40 per cent sand.

Soil fractions fall into four separate and distinct parts:

- the gravel fraction which can occur in six different shapes: rounded, irregular, flaky, angular, elongated, or elongated and flaky;

- the sand fraction (fine aggregate fraction of a soil) can be subdivided into four main zones - one to four - in ascending order of fineness. The zone number is determined by the amount of fine particles passing a 0.6 mm sieve;

- the silt fraction generally consists of fine ground rock which will hold together when damp and compressed. Too much water may make the soil spongy, but not sticky. Therefore careful analysis must be performed before it can be decided whether such soil can be used in block making; and

- the clay fraction which is further described below.[1]

The clay fraction is of major importance in the study of soil stabilisation because of its ability to provide cohesion within a soil. Mineralogically, clay may contain a variety of components such as kaolinite, vermiculite, illite, chlorite and montmorillonite. Clay minerals usually impart plasticity to the clays. Montmorillonite is extremely plastic and sticky, while kaolin is less so, and chlorites and vermiculites not at all.

---

[1] The composition of clays is described in detail in Grimshaw, 1971.

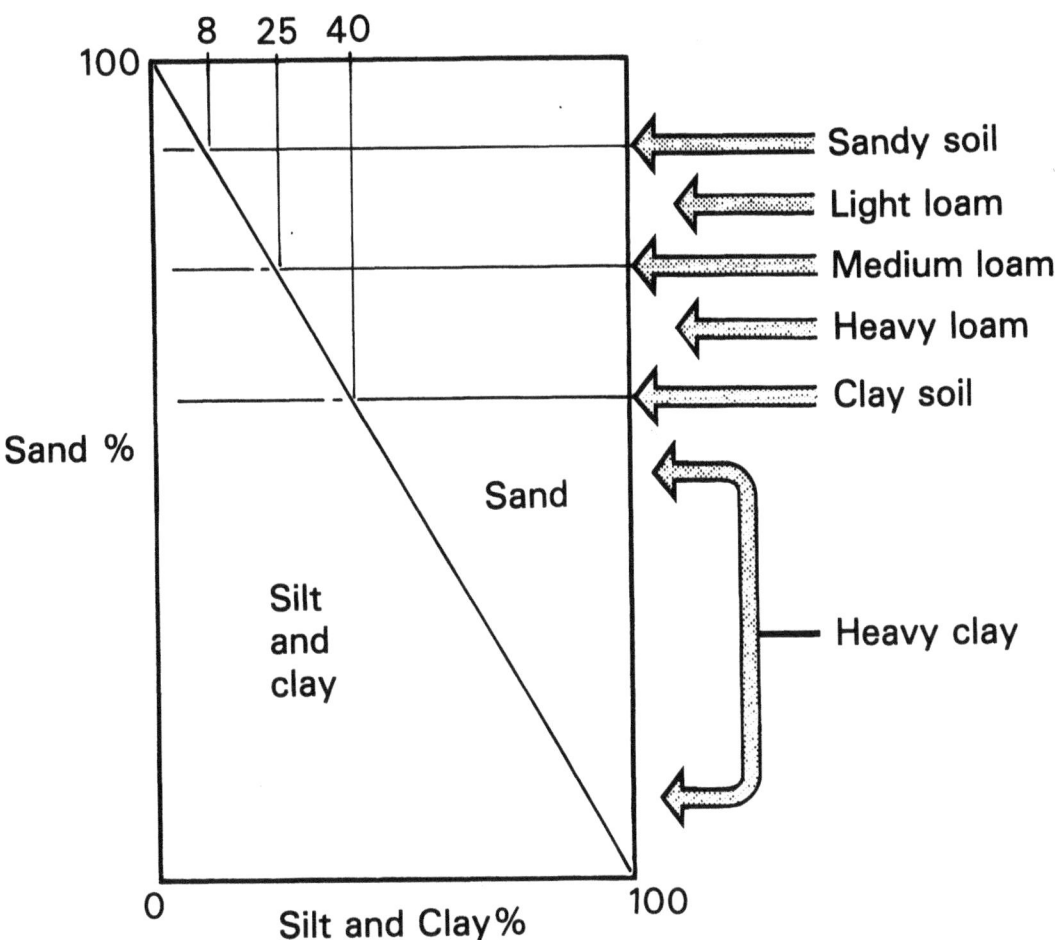

Figure II.1
Soil composition

Sand with silt and clay proportions

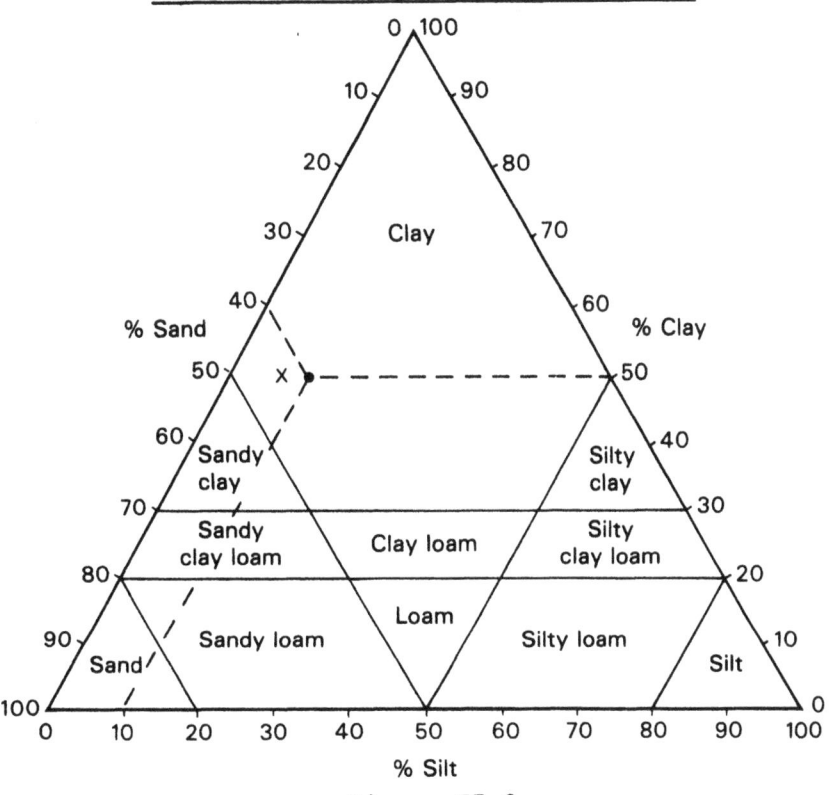

Figure II.2
Soil classification triangle

Kaolinite and montmorillonite represent opposite ends of the spectrum of the clay fractions. They differ in their ability to expand and contract when subjected to changing moisture conditions. For example, a typical black cotton soil from the Sudan having a combined silt and clay fraction of about 55 per cent cent, with the clay fraction containing montmorillonite, has a linear drying shrinkage of about 18 per cent. This type of soil also expands a great deal when moistened. On the other hand, a laterite soil with a predominance of kaolinite, has a low level of linear shrinkage.[1]

The production of good quality, durable stabilised soil blocks requires the use of soil containing fine gravel and sand for the body of the block, together with silt and clay to bind the sand particles together. A suitable type of stabilising agent must also be added to minimise the linear expansion that occurs when water is added to the clay fraction. The stabilising agent has other beneficial aspects which are described in a later section.

II. QUARRYING THE RAW MATERIAL

For small-scale, on-site manufacture of stabilised soil building blocks, a minimum of 700 tonnes per year of suitable soil is required for each block making machine.

The quarry should be as close as possible to the manufacturing site in order to minimise the trouble and expense of transporting the raw material. Sufficient soil must be available from the quarry site to meet the required scale of production.

Test holes

Trial holes must always be dug before major excavation commences to test the suitability of the soil and estimate available quantities. A cross section of the soil layers and zones, known as the soil profile, is illustrated in figure II.3.

The top soil (zone 1), usually dark in colour, contains fibrous materials and rotting vegetation; the lower layers of this zone may smell when wet and be very friable when dry.

---

[1] These characteristics are described in detail in Prescott and Pendleton, 1966.

Zone 2 soil should have a beige colour and will be very sticky if it contains a high clay fraction. Under wet conditions, clay soils will induce the formation of puddles of water and will be slippery and greasy to the touch.

The sandy soil (found in zone 3) is much easier to excavate, will not retain any free water and will feel gritty to the touch.

Several test holes should be dug close to one another. It is advisable to excavate a minimum amount of soil: a 15 cm diameter hole, 2 to 3 metres deep should usually be sufficient to obtain a full soil profile and detailed analysis of the clay and sand fractions.

Soils can vary widely even within a small area. For this reason, one should not be satisfied with what is found in a single test hole and should instead dig several holes in an area big enough to supply all of the soil that is needed. The number of holes to be dug must be determined in each case. Test holes are made according to the following steps.

One square metre of top soil should first be removed with a spade in order to expose the zone 2 soil layer. The depth of the top soil, which may vary between 15 cm to several metres should be recorded for future reference.

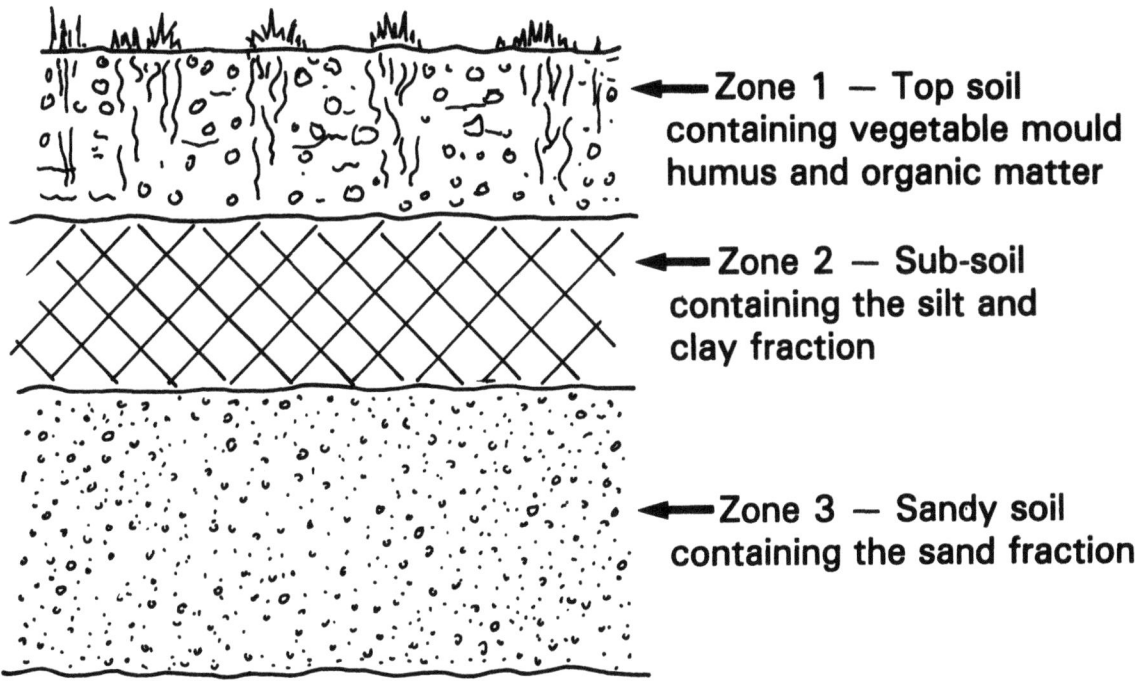

Figure II.3
Soil profile

The further excavation of a small diameter test hole is best achieved with a screw auger or bucket auger which are normally operated by two men. Figures II.4 and II.5 illustrate these two types of hand-operated, soil drilling equipment. Each of these tools can be fitted with varying lengths of screwed tubes to allow excavation of different depths. The operators must apply vertical pressure to the auger head via the screwed tube at the same time as rotating the cross handle.

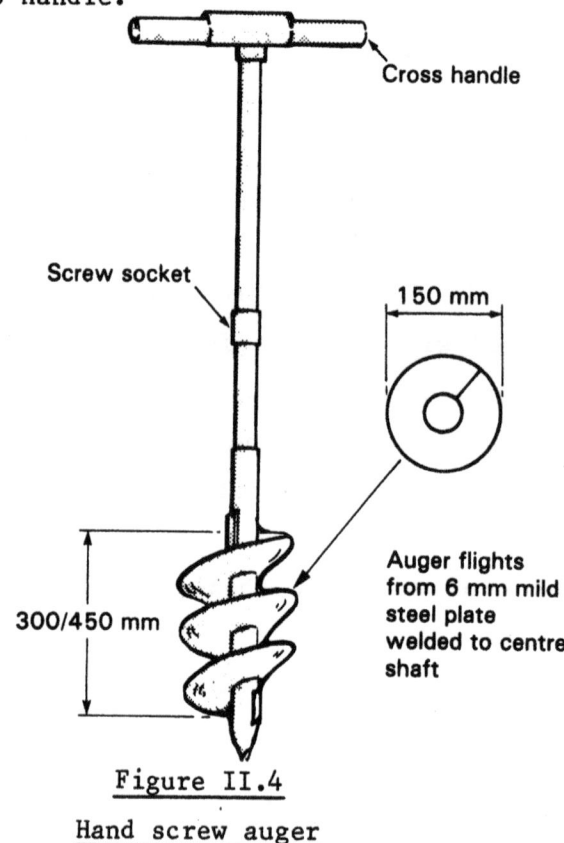

Figure II.4

Hand screw auger

When in use, the screw auger is rotated into the ground to a depth of about 20 cm, then lifted out, and the soil removed from the cutting blade flights. The bucket auger collects the excavated soil within its bucket-shaped flights and is emptied after removal from the ground. A hole of about 15 cm diameter is cut with the screw auger, whereas the smallest bucket auger produces a hole of about 25 cm diameter.

Whatever the type of auger used, an accurate depth record of soil conditions must be kept, along with a site-plan view of the location of the test holes. An example of such a site-plan is shown in figure II.6.

The screw auger can be manufactured locally in a blacksmith's shop by first cutting annular rings from 6 mm thick mild steel plate. These rings are then opened up to form the auger flights and welded to a centre shaft. The bucket auger (figure II.5) is more difficult to manufacture locally.

Quarrying equipment and tools

Different types of excavating tools can be used in a quarry, depending on the size of the proposed project. For a large project, a bulldozer can be brought on site to remove the top soil quickly (zone 1 in figure II.3). It is recommended that this top soil should be stockpiled so that it can be replaced and re-used for agricultural purposes after excavation. Excavation of zone 2 or 3 (see figure II.3) may require a mechanical drag line shovel (figure II.7).

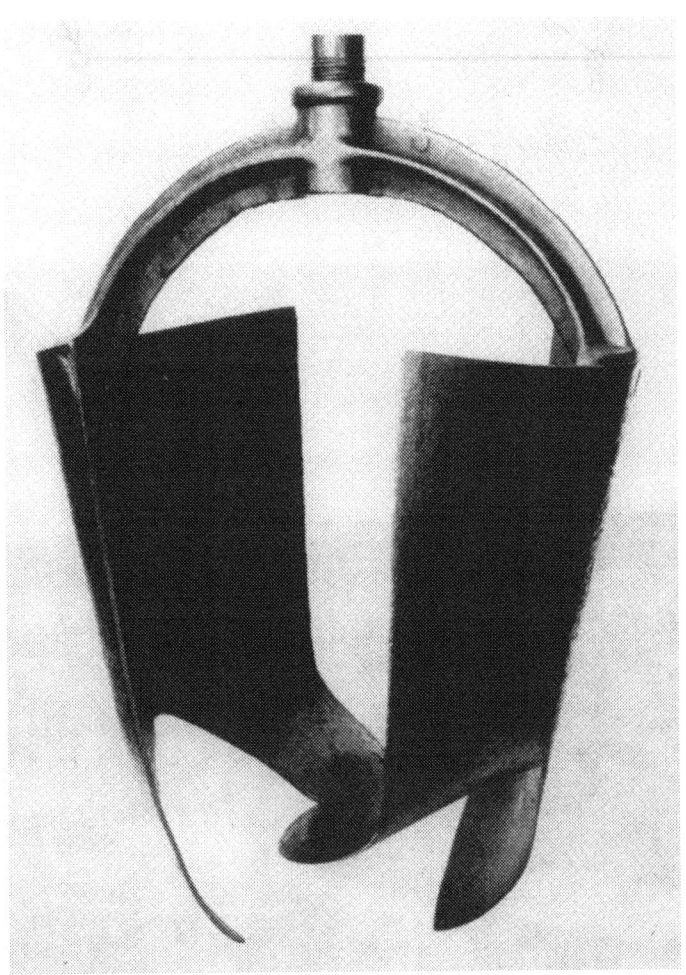

Figure II.5
Bucket auger

In view of the scales of production covered by this memorandum (up to a daily output of 400 blocks per block making machine), it is more economical to use wheelbarrows and the various hand tools available on the market. Hand digging has been found to be reasonably efficient even for medium-size brick works producing up to 10,000 fired bricks per day.

- 20 -

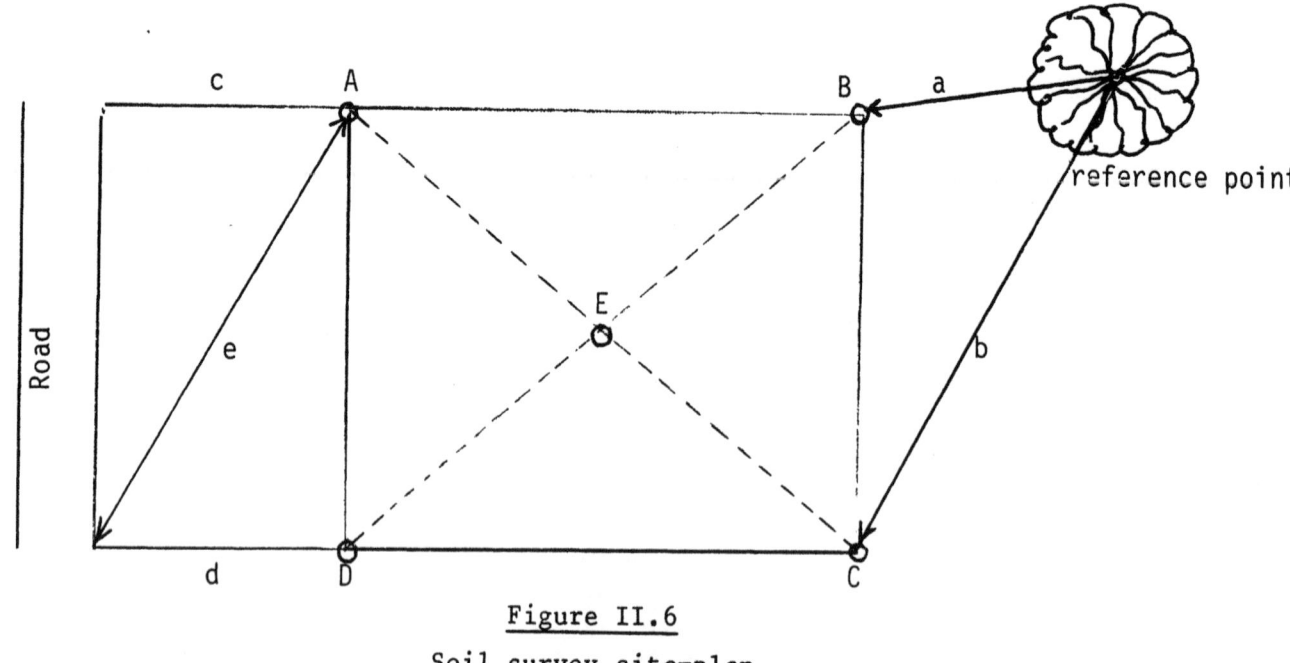

Figure II.6
Soil survey site-plan

Figure II.7
Mechanical drag line shovel

A major advantage of hand digging over mechanised excavation is that unwanted materials (e.g. large rocks and stones, uncrushable objects, tree roots) can be easily discarded when excavating. This is not easily achieved with mechanised excavation.

The spade or shovel is the most common type of handtool used for digging. The most common types are illustrated in figure II.8.

Type (a): a clay digging spade with a slightly rounded blade which can be used to dig both clay and heavy loams;

Type (b): a spade with a square-ended blade suitable for cutting through fibrous materials and skimming weed growth (e.g. top soil growth and grass);

Type (c): a builder's shovel, with upturned edges to prevent spillage; this is a very efficient handtool, ideal for general lifting and mixing duties;

Type (d): this type of shovel is slightly curved and has a pointed cutting edge; it was originally developed to handle asphalt; it is used in the building industry, although it is not very efficient for digging or mixing materials together; and

Type (e): a pick-hoe which has many uses for digging, breaking up hard ground and lumps; it is very efficient for both excavating and mixing duties.

Spades or shovels with shafts of different lengths and blades of different sizes are widely available. The standard shape of a spade is 29 cm long and 19 cm wide, whereas the shovel blade is 29 cm long and 24 cm wide.

The shafts of spades and shovels should have a gentle crank just above the point where they are joined to the blade to allow easier use and maximum leverage. The strapped or tubular socket should be securely attached to the shaft. Metal treads welded to the upper edge of the blade makes digging, especially that of heavy soils, less painful to the foot.

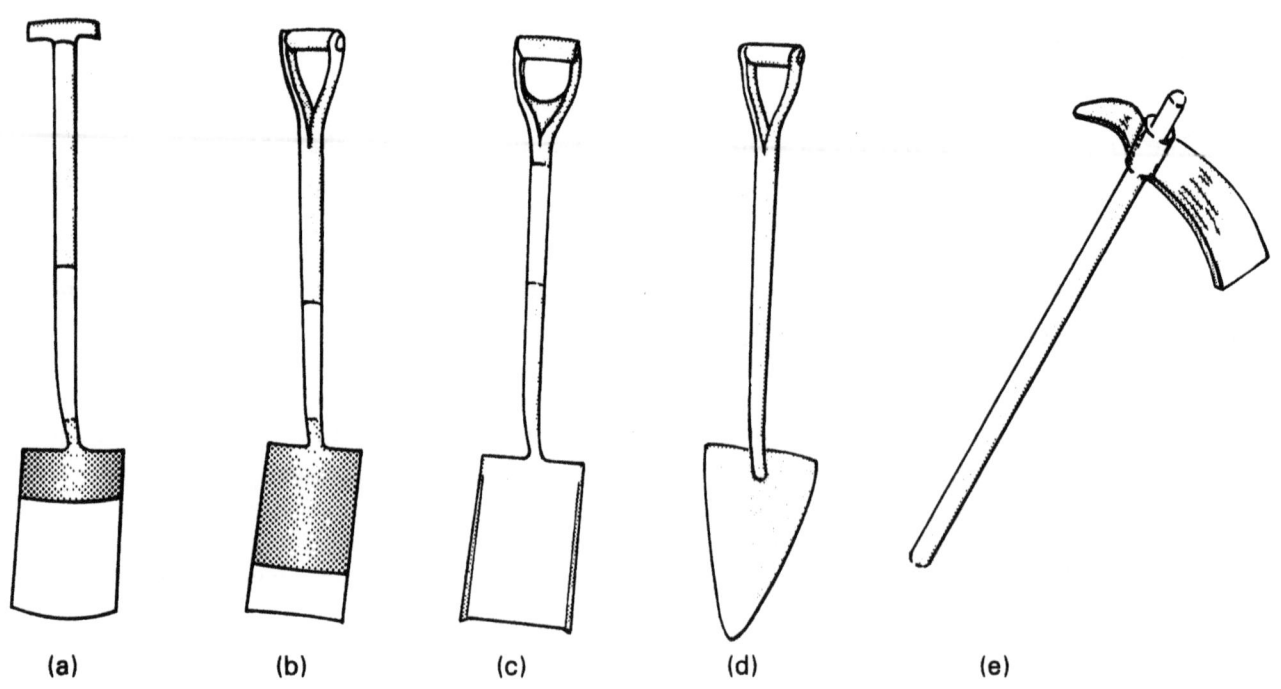

Figure II.8
Hand digging tools

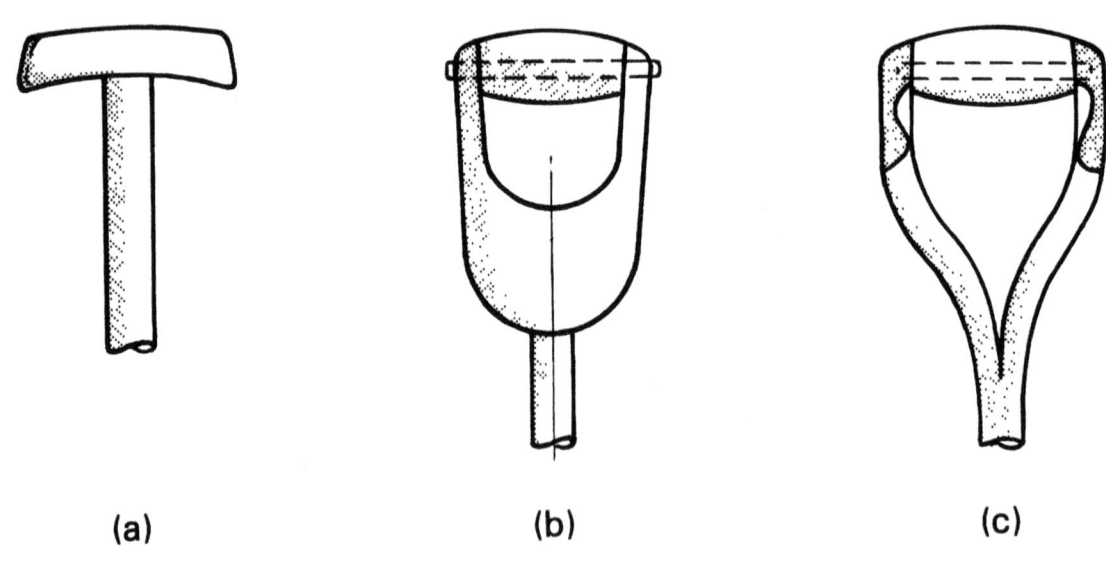

Figure II.9
Handle shapes

The spades and shovels illustrated in figure II.8 show three different types of handle shapes. These are illustrated in more detail in figure II.9. They are:

-type (a): A T-shape handle is less expensive but does not offer the same control as either type (b) or (c); this type of handle can easily be broken;

-type (b): A D-shape handle allows full control of the spade or shovel but has a limited life because the handle joints are exposed to water which can cause premature rotting and splitting of the hold piece; in addition, the steel assembly pin might corrode, become weaker and split the wood.

-type (c): A 'YD'-shape handle is the most comfortable shape to hold, being slightly larger than the 'D'-shape. It affords greater control and is therefore most efficient to use; it is, however, the most expensive of the three types; shaped metal shields are employed to protect the assembly joints.

III. SOIL TESTING PROCEDURES

A detailed investigation of the raw material is always desirable and a thorough laboratory analysis should always be carried out for large-scale production. It is not essential, however, to use sophisticated tests to determine the suitability of a soil for small-scale production. Simple preliminary tests can be conducted on site to obtain an indication of the components of a soil sample, its silt/clay and sand fractions, and to investigate soil mouldability, an essential characteristic in the manufacturing of stabilised soil blocks.

For soils which appear to be suitable at first sight, further tests should be carried out to determine the nature of the soil and to select a suitable stabilisation procedure.

III.1 Preliminary on-site tests

Soil samples from zone 2 and zone 3 soils (obtained from test holes) should be tested in the way described below:

<u>Smell test</u>: Damp soil emitting a musty odour indicates the presence of organic material and is therefore not suitable for block making. Such soil should be discarded.

Colour appearance: The dark brown crumbly humus in the soil is organic matter. Soil of this colour should in general be discarded. Light brown to black colouring indicates that the soil contains at least a small proportion of organic matter but that it may be suitable for stabilising. The colour test does not, however, work in all cases. For example, black cotton soils are dark brown to black in colour but do not contain much organic material.

A reddish to dark brown colour indicates the presence of iron oxides which are acceptable for soil stabilisation purposes. White to yellow colouring is an indication of the predominance of lime-based compounds or sand. This type of soil can be stabilised.

Pale brown colouring is characteristic of the presence of clay; lime might be needed as a stabilising agent for this type of soil.

Shine test: A small piece of dry soil is rubbed with the back of a finger nail in order to identify the main component in the sample. The soil surface is abrasive to the touch and the soil remains dull if sand or silt is predominantly present. On the other hand, a sample containing clay shines and is smooth to the touch.

Thread rolling test: This test requires adding sufficient water to a small quantity of soil so that the sample can be easily moulded by hand. The soil sample is then rolled out on a flat clean surface into a thread with the palm of the hand or the fingers (see figure II.10). The reduction of the thread to about 3mm in diameter indicates the presence of a high clay fraction. On the other hand, the breaking of the thread at a larger diameter indicates the presence of a moderate sand fraction. This test is also used to determine the plastic limit of a soil (see section III.3).

Hand moulding test: After having removed stones and any foreign bodies larger than about 6 mm diameter, the soil sample is moistened and formed into a cube with an edge of about 2.5 cm. If a cube is formed easily, a high clay fraction is present. Although good adhesion and mouldability of such soil are advantageous in the block making process, too much clay will make the soil sticky to work with, and its high shrinkage may lead to cracks within the manufactured soil blocks.

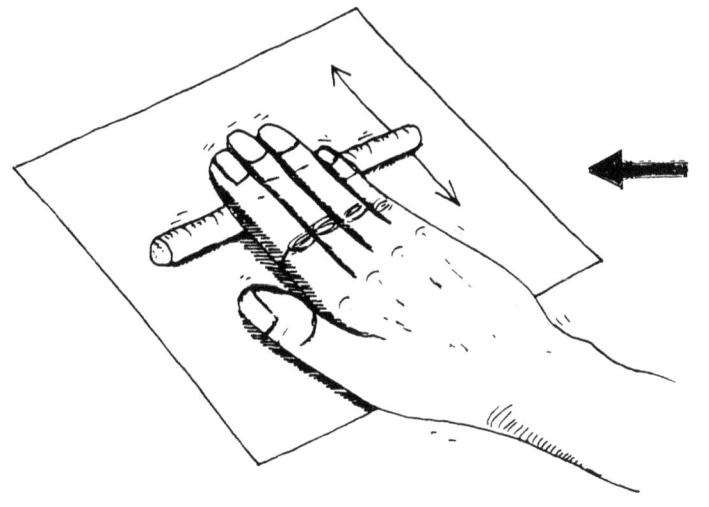
Sample rolled out on a flat clean surface (glass, marble, metal)

Reduction of thread to about 3 mm diameter: high clay fraction

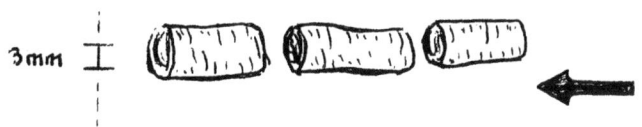
Breaking of the thread at a larger diameter: presence of a moderate sand fraction

Figure II.10
Thread rolling test

Next, the moulded test "cube" is allowed to dry out in the sun for one day. The occurrence of any surface cracks indicates a high clay fraction, which may give similar cracking problems in the blocks. On the other hand, the splitting of the cube into several pieces indicates the presence of too much sand or silt. Blocks produced from such soil may also fall apart.

III.2 Further soil testing procedures

The preliminary on-site testing methods described above will indicate whether a soil is likely to be suitable for stabilised soil block production. These tests may not, however, be sufficient. Other tests may be necessary, especially if the preliminary tests are not conclusive.

Sophisticated laboratory methods of soil testing, including chemical and sieve analysis and determination of the plastic limit, liquid limit and the optimum moisture content for maximum soil density have all been evolved by soil engineers. However, these laboratory tests are expensive and time-consuming and are only deemed necessary for large-scale projects. For a small project, fairly effective but simple on-site tests requiring simple equipment which may be locally manufactured can be conducted.

After preliminary on-site tests on soil samples obtained from test holes, the holes producing *a priori* good quality soil should be opened up in order to collect a larger sample for more detailed examination. The following on-site tests may then be performed:

Particle size distribution: This test gives a quantitative measure of the individual soil fractions. It requires four sieves and a tray similar to those illustrated in figure II.11; these sieves nest onto one another for proper site sieve analysis.

The four sieves must have different wire mesh sizes (e.g. 6 mm, 2 mm, 0.2 mm and 0.06 mm). The 0.06 mm mesh may be difficult to obtain and could be replaced by an open weave cloth. The fifth container is a catchment tray. The test should be performed according to the steps noted below.

A sun-dried soil sample of 2 kg is first weighed out and placed inside the 6 mm sieve located on top of the nest of sieves. By shaking the nest of sieves simultaneously, all the fine particles pass through this sieve and,

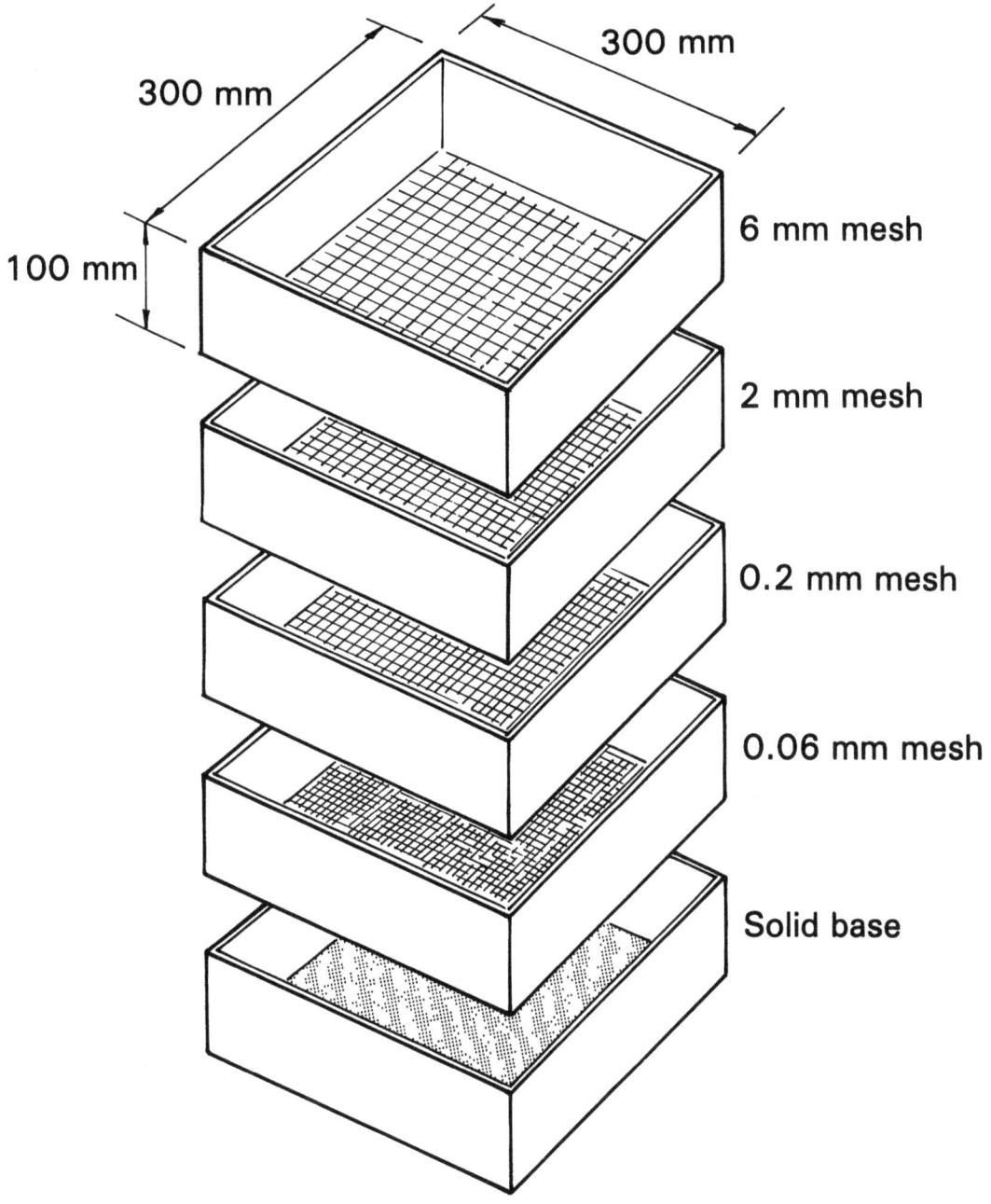

Figure II.11
Site sieves

depending on their fineness, some will rest on intermediate sieves, while those passing the 0.06 mm sieve will fall into the catchment tray.

Once the transfer of material from one sieve to another has ceased, the separated fractions of soil lying on top of each sieve and in the catchment tray are removed, weighed and recorded. A simple particle size distribution is thus obtained for soil sampling.

The fraction of soil retained on the sieves may be classified as follows:

Sieve mesh size	Designation of the fraction retained on the sieve
6 mm	Coarse and medium gravel
2 mm	Fine gravel
0.2 mm	Coarse and medium sand
0.06 mm	Fine sand
Catchment tray	Combined silt and clay

The results of the sieve analysis give an indication of the type of stabilising agent best suited for the soil. Ideally, there should be an even distribution of each soil fraction in order to manufacture good-quality stabilised soil building blocks. If this were to be the case, about five per cent cement would be needed as a stabilising agent. In practice, it is generally found that one fraction is larger than the others. For example, if there is a high fraction of coarse and medium sand and a low silt/clay fraction (e.g. less than about 20 per cent), about four to six per cent cement should be used to stabilise the soil. Conversely, if the silt/clay fraction is high, (e.g. above about 30 per cent), about six to eight per cent lime can be used as a stabilising agent. However, there may be a high proportion of silt present which would affect the linear shrinkage properties of soil; in this case, cement may be required.

Sedimentation bottle test: This test gives more information on the finest particles contained within a soil sample. It is performed in the manner noted below.

A wide-necked, straight-sided and flat-bottomed bottle or jar is needed for this test. The bottle is first filled to one-third with clean, uncontaminated water (see figure II.12(1)). Approximately the same volume of dry soil (which has passed through the 6 mm sieve) and a teaspoonful of common salt are added. Salt facilitates the dispersion of soil particles (see figure II.12(2)).

The lid is then firmly fixed on the bottle and the contents well shaken. When the soil and water have been mixed, the bottle is placed on a flat surface for about half an hour. Then, the bottle should be shaken again for two minutes and replaced on the level surface. Two or three minutes later, the water will start clearing. The finer particles fall more slowly and are thus deposited on top of the larger size particles. Two or three distinct layers will be observed, with the lowest layer containing fine gravel, the central layer containing the sand fraction and the top layer containing the combined silt and clay fraction. Figure II.12(3) illustrates this layer formation in a bottle. The individual percentages can be determined by direct measurement of the depth of each layer.

<u>Linear shrinkage mould test</u>: This test indicates the linear shrinkage of a soil sample as it dries. This information will help determine the best type and amount of stabiliser required. This test requires first the construction of a linear shrinkage mould with the following internal dimensions: 40 mm x 40 mm x 600 mm. Figure II.13 illustrates the mould required together with leading dimensions.

The first step in this test is to lubricate the internal faces of the mould with some type of oil or grease. Ideally, silicone grease is preferred but any type of mould release oil or grease could be used. The lubricant reduces soil drag on the internal faces of the mould occurring as the soil sample dries out and shrinks.

The soil sample which passed through the 6 mm sieve is mixed with water until a wet puddingy mix is obtained (this occurs near the liquid limit - see section III.3). This mix is then packed into the mould cavity, ensuring that the mould is completely full (absence of air pockets) and the top open surface is smooth. The mould is then placed to dry either in the sun for about five days or under shading for about ten days. In either case, it must be protected from rain.

If the soil has a high clay content, the sample will shrink and hog up out of the mould. This is illustrated in figure II.14 which shows the

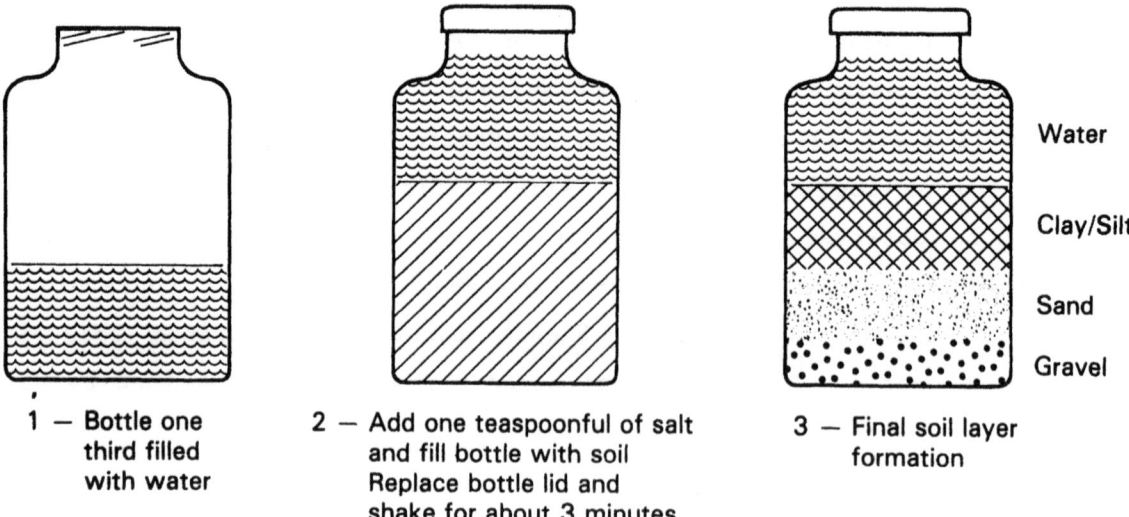

Figure II.12

Sedimentation bottle test

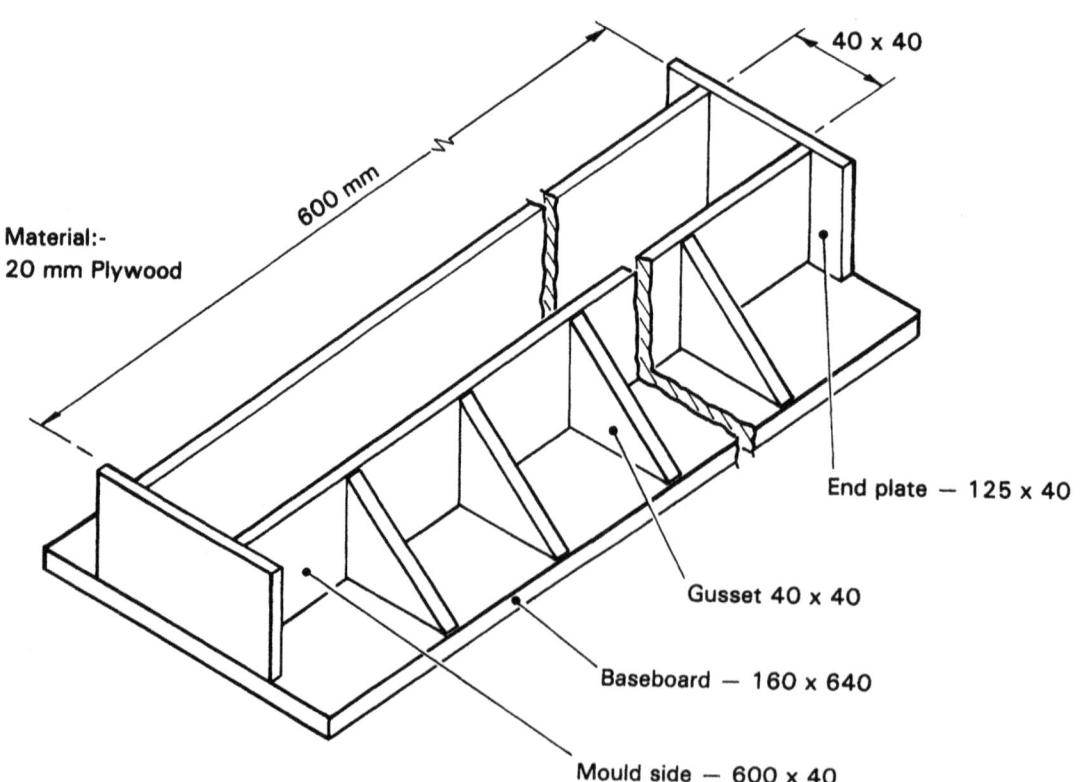

Figure II.13

Drawing of linear shrinkage mould

shrinkage properties of black cotton soil. A soil sample which shrinks and cracks across the width of the mould (see figure II.15) indicates a high sand fraction and low silt and clay fractions.

The linear shrinkage can be determined by subtracting the length of the dry soil sample from the length of the mould cavity. This shrinkage is usually expressed as a percentage of the original mould cavity length.

III.3 Laboratory testing methods

Until 1939, the science of soil mechanics was almost entirely in the research stage, and with the exception of the liquid limits, there were no standard tests to determine the engineering properties of a soil. Since then, increased knowledge of soil properties and its frequent use in practical engineering has led to a convergence of soil testing methods used in different countries, and to the formulation of national standards.

A large number of simple or sophisticated laboratory tests are currently used in various countries.[1] However, the following laboratory tests should be sufficient for assessing materials for the production of stabilised soil building blocks. These tests are briefly discussed below.

Optimum moisture content (OMC): This characteristic of soils is defined[2] as the moisture or water content at which a specified amount of compaction will produce the maximum dry density. With relation to soil, a low moisture content will affect the extent to which the soil can be compacted under pressure. In this case, individual soil particles cannot come into close contact with one another, thus allowing the presence of some air spaces between them. If, on the other hand, the moisture content of a soil is high, there will be a greater flow of particles when pressure is applied but these particles will be separated by a film of moisture. Ultimately, as the soil dries, the water evaporates, leaving air spaces between the particles. Consequently, high and low moisture contents will result in poor compaction, which is synonymous with low density. The relationship between dry density and percentage moisture content is illustrated in figure II.16.

---

[1] Some of these tests are described in Akroyd, 1962.

[2] The definition may be found in British Standards Institution, BS924, 1975.

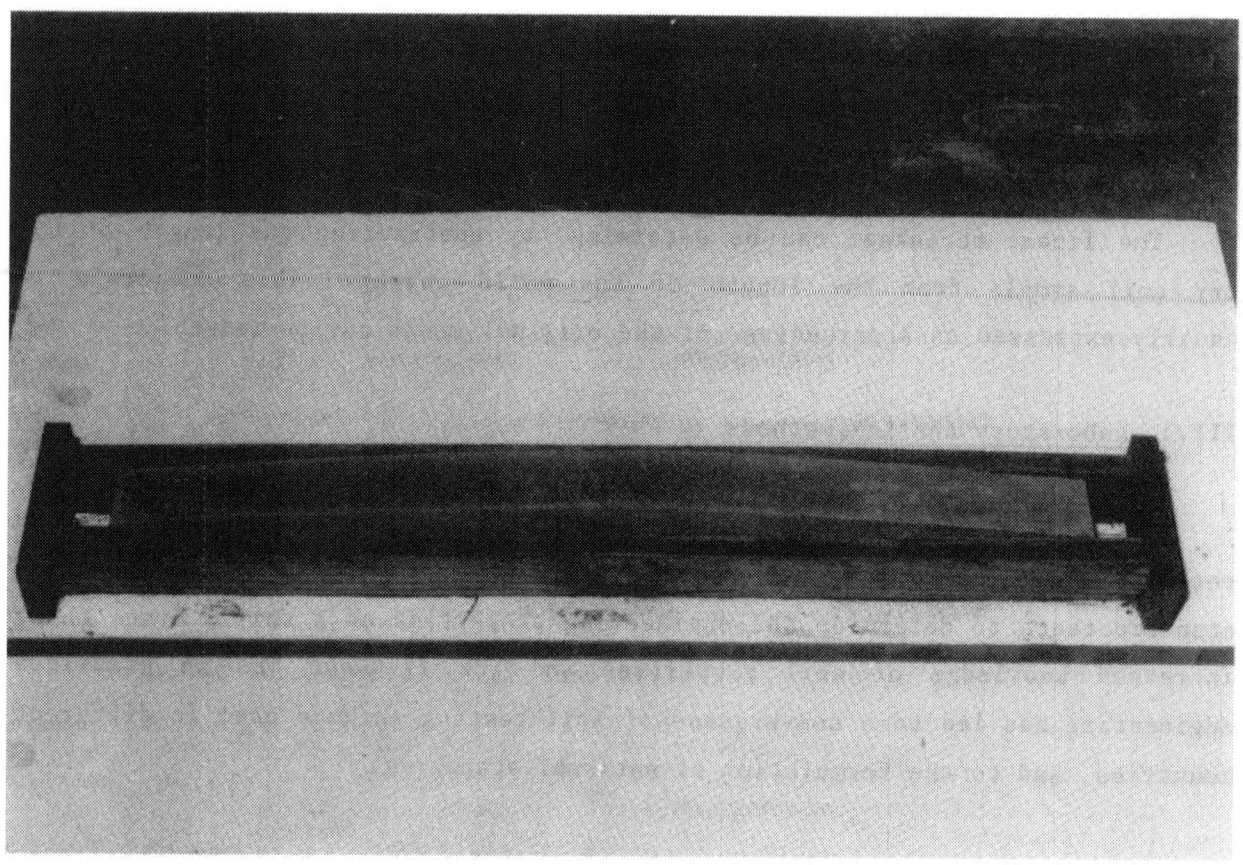

Figure II.14
Linear shrinkage of a high silt/clay soil

Figure II.15
Linear shrinkage of a sandy soil

A compromise can be found between extremes of moisture content to minimise air voids and therefore to obtain maximum compaction and density. The moisture content corresponding to the highest dry density is defined as the optimum moisture content.

It may be shown that the OMC and the density of a soil depend upon the type and quantity of stabilising agent employed and the method of compaction used.[1] Therefore, the optimum moisture content should be determined on the basis of a prior knowledge of the type and quantity of stabilising agent which is intended to be used for a given amount of soil and of the selected compaction method.

<u>Liquid limit</u> (LL): The liquid limit is defined as the moisture content at which a soil passes from the plastic to the liquid state. The method employed to determine the liquid limit consists first of placing a soil-water paste in a standard cup. The paste is then divided into two halves with a grooving tool. The moisture content at which the two halves will flow together when the cup is given a standard number of blows is finally determined. This moisture content corresponds to the liquid limit of the mixture.

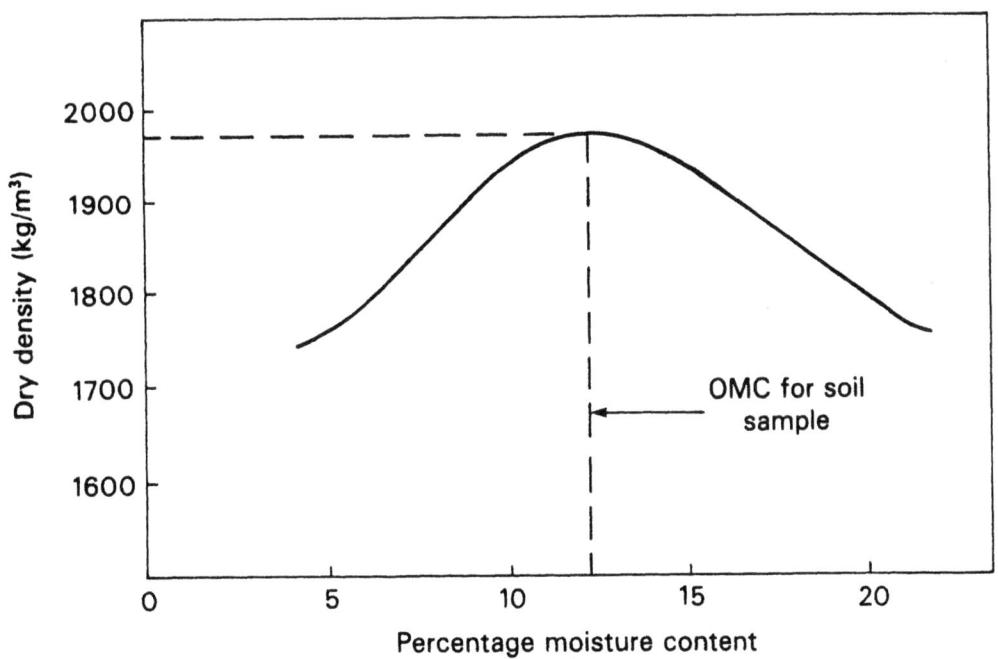

Figure II.16
<u>Typical density/moisture curve</u>

---

[1] See Lunt, 1980.

Plastic limit (PL): The plastic limit is defined as the moisture content at which the soil becomes too dry to be in a plastic condition. The plastic limit is determined by rolling a thread of soil to 3 mm in diameter between the fingers and a glass plate. The soil will be at its plastic limit if the thread just crumbles under this rolling action.

Plasticity index (PI): The plasticity index is defined as the numerical difference between the liquid limit and the plastic limit:

$$PI = LL - PL.$$

Particle size distribution: This test relates to the quantitative determination of the particle size distribution in a soil down to the fine sand fraction. The combined silt and clay fraction can be obtained by a wet sieving method[1] or by a pipette method to determine the individual silt and clay fractions. The procedure involves the preparation of a soil sample by wet sieving to remove the silt and clay fractions, followed by dry sieving of the remaining coarser material.

Chemical tests: There are two distinct chemical tests employed to check the suitability of a soil:

- determination of the organic matter content; and
- soil chemical analysis.

These two tests are briefly described below.

Organic matter testing: Organic matter takes the form of humus which usually occurs in the top soil layer or zone 1. This organic matter will seriously impair the setting or hardening of cement or will affect the pozzolanic reaction between hydrated lime and the stabilisation of the soil.

The best method to check the presence of organic impurities consists in determining the pH value of a soil (i.e. the level of acidity or alkalinity of a compound[2]). The pH of a soil sample is determined in the following manner. The sample is shaken vigorously with excess distilled water in a

---

[1] This method is described in West and Dumbleton, 1972.

[2] The testing method for determining the presence of organic materials is described in the British Standards Institution BS1924, 1975.

glass container and allowed to settle. A chemical indicator is then added to the supernatant water. The resulting change of colour of the indicator indicates the pH of the soil. The following colour changes indicate the degree of acidity or alkalinity of a sample:

- red: high degree of acidity (pH lower than 5,5);
- orange to yellow: low degree of acidity (pH between 5.5 and 6.5);
- brownish: neutral sample (pH between 6.5 and 7.0);
- green to green-blue: low alkalinity (pH between 7 and 8);
- blue: high degree of alkalinity (pH greater than 8).

Soils with pH readings above 10 and below 4.5 are rare. They should not be used for soil stabilisation projects because they have high impurity levels. Their use requires high proportions of stabiliser and will therefore considerably increase production costs.

Chemical analysis: The chief purpose of a full chemical analysis is to identify all the elements present and their proportions. In some instances, it may reveal the presence of an unsuspected mineral which might affect the stabilisation process. The results can also be used to determine whether the soil can be classified as a true laterite, a lateritic or a non-lateritic soil.

Table II.2 provides the percentage of various chemical compounds present in soil samples from four countries. The following remarks can be made regarding the suitability of these soils for block making:

- the sum of the fractions of alumina, silica and iron oxide must be greater than 75 per cent; this is the case for the four soil samples;

- the percentage loss on ignition (LOI) must be less than 12 per cent. Higher figures will indicate the presence of organic matter which would affect the hardening of a stabilised soil block; thus, the Kenya soil sample would be suspect and might not be found suitable for a soil stabilisation project;

- soluble salts in a clay may influence the plasticity of the soil and will affect the long term strength of a stabilised soil block; these salts are often compounds of potassium and sodium; a combination of potassium and sodium oxides greater than 2 per cent constitutes an undesirable amount of soluble salts; thus, the Egyptian soil sample would be suspect.

Table II.2

Chemical soil analysis

(percentage)

Chemical component	Chemical symbol	Jamaica red	Kenya red coffee	Sudan black cotton	Egypt
Alumina	Al_2O_3	17.20	32.90	9.18	18.30
Silica	SiO_2	62.50	36.20	76.80	51.30
Phosphorus pentoxide	P_2O_5	0.01	0.22	0.01	0.15
Sulphur trioxide	SO_3	0.01	0.01	0.01	0.83
Potassium oxide	K_2O	0.25	0.36	0.45	1.17
Calcium oxide	CaO	0.35	0.41	1.85	2.59
Titania	TiO_2	0.93	1.52	0.68	0.98
Manganese oxide	Mn_2O_3	0.04	0.33	0.05	0.05
Iron oxide	Fe_2O_3	8.39	10.72	3.54	8.19
Sodium oxide	Na_2O	1.13	0.27	0.33	3.32
Magnesia	MgO	0.55	0.24	0.46	1.79
Loss on ignition	LOI	9.40	18.10	6.24	11.66

- the four soil samples may be classified as lateritic or non-lateritic soils according to the value of the following ratio:[1]

$$\frac{\text{Percentage of silica}}{\text{Sum of percentages of alumina and iron oxide}}$$

The following table indicates the classification of soils according to the value of the above ratio:

Soil types	Value of ratio
Laterite	1.33 or less
Lateritic soil	1.33 to 2.0
Non-lateritic soil	2.0 and above

The four soil samples from table II.2 may thus be classified as follows:

---

[1] Most soil engineers, chemists and geologists working in the field of soil stabilisation use this method of soil classification.

Kenya sample: true laterite
Egypt sample: lateritic soil
Jamaica sample: non-lateritic soil
Sudan sample: non-lateritic soil

IV. SOIL STABILISERS

Methods to improve the natural durability and strength of a soil - commonly referred to as soil stabilisation - are practised in many countries. These methods are not new, since stabilisers (e.g. natural oils, plant juices, animal dung and crushed ant hill materials) have been used for many centuries. In recent years, scientific rather than ad hoc techniques of soil stabilisation have also been introduced, developed largely from early methods devised for the stabilisation of earth roads.

IV.1 Principles of soil stabilisation

The silt and clay fraction of a soil reacts to the application of water, swelling when taking in water and shrinking on drying out. This movement can produce cracking of walls and accelerate erosion, which, if serious, may lead to structural failures. Furthermore, the movement often causes the crumbling of protective renderings which may have been applied to the surface of the wall.

The aim of soil stabilisation is to increase the soil resistance to the erosive effects of local weather conditions, including changes in the temperature, humidity and rain.

A better soil resistance to erosion can be achieved in one or more of the following ways:

- by increasing the density of a soil;
- by adding a stabilising agent that either reacts with or cements the soil particles together; and
- by adding a stabilising agent which acts as a waterproofing agent.

The use of the correct stabilisation method might improve the compressive strength of a soil by as much as 400 to 500 per cent and increase its resistance to erosion.

IV.2 Soil stabilisation methods

There are seven main methods of soil stabilisation. These are described and assessed in this section.

(i) <u>Manual or mechanical stabilisation method</u>: This method increases, through mechanical means, the density of a soil and therefore improves its durability. The easiest way of increasing soil density is to ram or tamp a slightly moistened soil mix in a mould in order to eliminate the air pockets;

It was shown in section III.3 that the highest block density may be achieved by compaction once the soil has reached an optimum moisture content. A standard test[1] may be used to determine the OMC value for a given type of soil.[2] The latter may then need to be moistened or dried in order to achieve this value before the soil can be used for block making. For example, with a compaction pressure of 3 MN/m^2 on a soil containing about 50 per cent silt and clay, a maximum dry density of 1980 kg/m^3 may be achieved with an OMC value of 12 per cent (see curve in figure II.16).

Manual compaction methods vary from foot treading to hand tamping equipment, with compacting pressures varying between 0.05 to about 4 MN/m^2. Mechanical equipment may achieve compacting pressures of several thousands MN/m^2. However, such equipment is outside the scope of this memorandum as it is not economically feasible for small-scale production.

(ii) <u>Cement stabilisation</u>: Ordinary Portland cement (OPC)[3] is the type of cement most widely used in the world today. It is made from a mixture of limestone and clay, heated to around 1,500°C. Gypsum is then added and the resulting mix ground to a fine powder. Portland cement hydrates when water is added and produces a cementitious compound independently of any aggregate.

When cement is added to a high-sand-fraction soil, the sand particles act as a filler. Thus, after the water is added to the mix, hydration occurs and the soil particles are embedded in a matrix of hard cementitious gel. The

---

[1] The British Standard Institution, BS1377, 1975.

[2] It may be noted that this value will generally change with the addition of a stabilising agent.

[3] For example, OPC manufactured to British Standard 12: see British Standards Institution BS12, 1971.

small proportion of lime released during the hydration process may react further with the small clay fraction of the soil mix, forming additional cementitious bonds within the soil-cement mix.

For effective stabilisation, it is important that the clay fraction is not so high as to swamp the small percentage of cement added to the soil mix. Therefore, it is necessary to increase the cement content of a soil mix as the clay fraction of a soil increases. The relationship between the linear shrinkage observed and the cement to soil ratio required has been established by the non-governmental organisation VITA.[1] Table II.3 shows that the cement to soil ratio varies between 5.56 per cent and 8.33 per cent as the measured shrinkage varies between 15 mm and 60 mm (by means of the shrinkage test).

Table II.3

Cement to soil ratio

Measured shrinkage (mm)	Cement to soil ratio
Under 15	1:18 parts (5.56 per cent)
15-30	1:16 parts (6.25 per cent)
30-45	1:14 parts (7.14 per cent)
45-60	1:12 parts (8.33 per cent)

It may be noted that, for a given shrinkage, cement to soil ratio is a function of the compacting pressure exerted. For example, a CINVA-Ram machine exerts a compacting pressure of about 2 MN/m^2 (see Chapter IV). If this pressure is increased to about 10 MN/m^2 (e.g using a different machine), the cement dosage could be reduced to between 4 and 6 per cent for soils with a shrinkage of up to 25 mm. Above this shrinkage value, 6 to 8 per cent lime (see below) could be used for effective stabilisation.

(iii) <u>Lime stabilisation</u>: The production of hydrated lime is carried out in two stages.

The first stage requires the calcination of limestone (or shells or coral) in a kiln at 900°C. This stage expels carbon dioxide and produces

---

[1] See Volunteers in Technical Assistance, 1977.

quick lime or calcium oxide. The second stage involves slaking or hydrating quick lime with a certain volume of water which causes the production of hydrated lime or calcium hydroxide.

Both quick and hydrated limes can be used to stabilise soils containing a high clay fraction.[1]

When lime is used as a stabiliser for soils with a high clay content, four reactions are supposed to occur:

- a cation exchange (a chemical exchange of ions takes place, giving the clay a lower affinity for water); the resulting mix is thus characterised by a lower moisture movement;
- flocculation or agglomeration follows as a result of the cation exchange; this results in the formation of clusters of the microscopically small soil particles, making the mix more viscous or stiff);
- carbonation of the lime itself, as it reacts with the carbon dioxide from the air, gives rise to a hardening effect; and
- a pozzolanic reaction (i.e. a chemical reaction between the clay and the lime, yielding hydrated calcium silicate aluminate compounds similar to some of those found in Portland cement). The rate at which this pozzolanic reaction proceeds is a function of the temperature. Thus, it is very low in temperate climates, but usually fast in the tropics.

The first two reactions take place as soon as the lime is added to the soil. The last two reactions are slower, causing the strength of lime stabilised soil blocks to develop over weeks, months or even years.

It has been suggested that when lime is used as a stabiliser instead of cement, the dosage should be double.[2] However, research at the United

---

[1] Lime is a caustic material that can cause damage to the eyes and skin. Careful handling is therefore advised, especially with quick lime which can react explosively if mixed incorrectly with water.

[2] See Volunteers in Technical Assistance, 1977.

Kingdom Building Research Establishment shows that such doubling is not necessary if a sufficiently high compacting pressure (e.g. a higher pressure than that provided by the CINVA-Ram press) is applied on a high clay content soil. Thus, the volume of air voids brings the lime and soil particles into closer contact, and the stabilising reactions can take place as fully as possible. For example, tests show that wet compressive strengths between 3.0 MN^2 and 3.5 MN/m^2 may be obtained with compacting pressures in the range of 8 to 14 MN/m^2. This is illustrated in figure II.17 with blocks made from Sudanese black cotton soil, tested over a wide range of compaction pressures. Eight per cent of lime is used as the stabilising agent with a soil which has a high silt and clay content of 58 per cent and a linear shrinkage of 11 per cent.

The main advantage of lime over Portland cement as a stabilising agent is that relatively simple equipment is required for its production, thus facilitating local manufacture. However, it has often been found that hydrated lime is more costly than Portland cement in countries where both materials are available. In rural areas, the difficulty of obtaining cement will often dictate the use of lime.

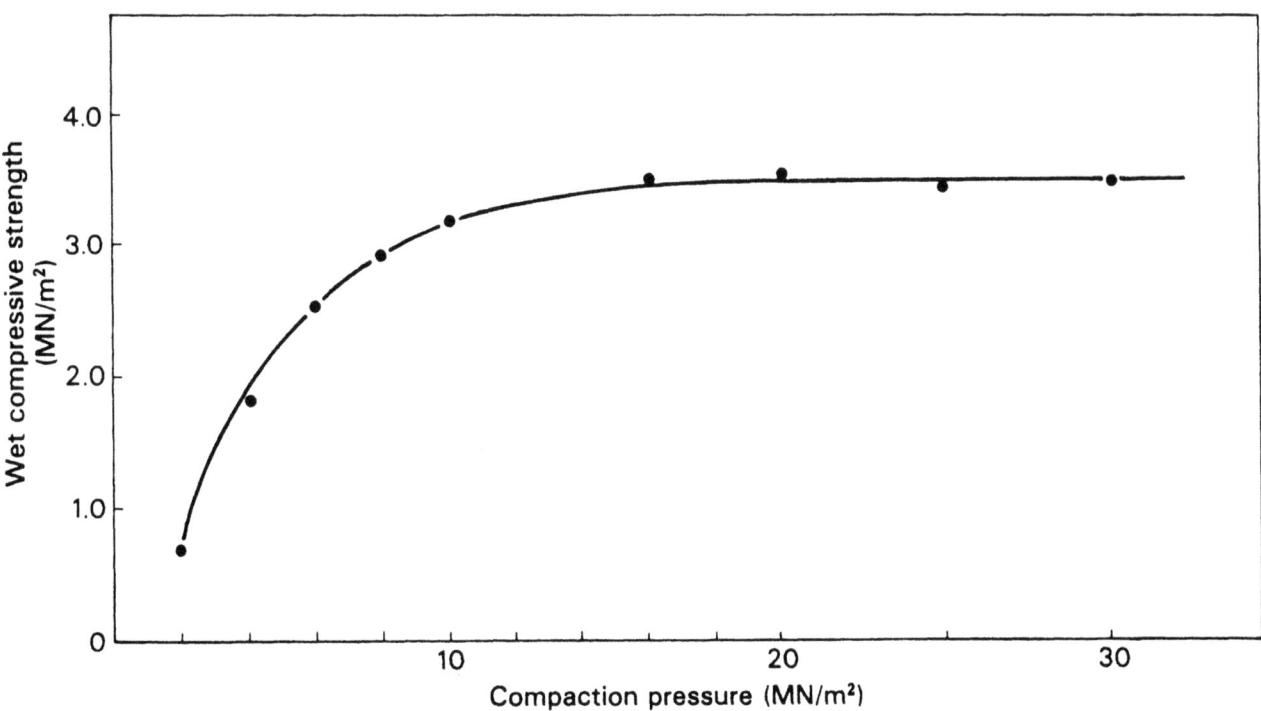

Figure II.17
Strength/compaction pressure curve of lime
stabilised soil building blocks

(iv) <u>Bitumen and bitumen emulsions</u>: In its natural form, bitumen or asphalt is too thick to be added to the soil. It is usually warmed to change it into a fluid and mixed with organic solvents, such as benzine, to make it thinner. It is emulsified with water for the production of a bitumen emulsion. This emulsion is mixed with a soil so that, when the moisture dries out, the bitumen reverts back to its natural state. This results in binding soil particles together. Little extra strength is gained by the soil. The main advantage of the operation is the waterproofing of the blocks which can then better withstand rain or humid weather conditions.

The most suitable soils for bituminous stabilisation are sands and sandy soils. Soils with a high clay fraction would require uneconomically large amounts of bituminous emulsion in order to obtain satisfactory results.

Stabilisation with a bituminous emulsion is not usually recommended because material costs are high. Furthermore, the heat of tropical sun tends to soften the block surface so that anyone touching the wall might get dirty from a bitumen deposit.

(v) <u>Gypsum plaster</u>: Gypsum plaster (or plaster of Paris) is produced by heating gypsum rock to about 170°C. At this temperature, 75 per cent of crystallisation water is driven off, leaving a white powder. The latter gets hard after mixing with water and settling over a short period of time. This material is usually employed for finishing internal wall surfaces and occasionally as a mortar. It is slightly soluble in water. Occasionally, gypsum plaster is used as a soil stabiliser for medium range clay-content soils. However, blocks made from such a mix are not very durable due to their low water resistance. They should therefore be used only for internal walls.

Gypsum plaster soil blocks were used in Australia for external walling. They required a protective covering or cladding of metal sheeting on the external faces of the walls. These protected gypsum plaster soil blocks developed sufficient strength to act as load bearing blockwork.

(vi) <u>Chemical stabilisers</u>: Different chemical compounds have been tested as stabilising agents. However, they require the application of sophisticated production techniques which are outside the scope of this memorandum.

(vii) <u>Other stabilisers</u>: Many so-called "stabilisers", such as animal dung ant heap material, bird droppings and animal blood, have been used for the manufacture of stabilised soil blocks. These waste materials generally contain nitrogenous organic compounds which, when wetted, form a gluey substance which helps bind together soil particles.

Chopped straw, grasses and natural organic fibres, although not active stabilisers, are used as reinforcement material to minimise linear shrinkage problems which occur with high clay content soil.

Agricultural waste, such as rice husk ash, cotton stalks, ash from burnt crushed sugar cane (bagasse ash), skimmed lime sludge from a sugar refining process (which burns spontaneously, leaving a black filter cake mud), resins and oils, are also used to a limited degree for soil stabilisation.

The above materials are often used in the production of sun-dried adobe blocks in rural areas. Although they provide only a small increase in strength to the blocks, they are a useful addition to a village-scale block production unit. Most of these "stabilisers" are readily available within a rural community.

CHAPTER III

PRE-PROCESSING OF RAW MATERIALS

I. <u>THE NEED FOR PRE-PROCESSING</u>

The raw materials used in the production of stabilised blocks are soil, stabiliser and water. The stabiliser, be it lime, ordinary Portland cement or some other material, is usually available in a powder or liquid form, ready for use. The soil may be wet or dry when it is first obtained, and it may not be homogenous. For example, it could contain stones or hard lumps. Inclusion of the latter would lead to poor quality products since both the rates and amounts of drying shrinkage of these inclusions differ from those of the main body of the material. These differences give rise to stresses in the blocks during drying, causing cracks and splits in the blocks which spoil their appearance and lower their strength and durability. To prevent this, it is often necessary to crush the soil so that it can pass through a 5 to 6 mm mesh sieve.

Different types of soils may also need to be used together in order to obtain good quality blocks. For example, a very sticky clay may be improved through the addition of sandy soil.

It is imperative not only to measure the optimum proportions of ingredients, but also to mix them thoroughly. Mixing brings the stabiliser and soil into intimate contact, thus increasing the effectiveness of physical processes, chemical reactions and cementing actions. It also reduces the risk of uneven distribution of the stabiliser in the soil and consequently the production of inferior quality blocks. Although heavy duty, large-capacity mixing machinery is available from manufacturers of clayworking equipment, it is too expensive and inappropriate for the type of production considered in this memorandum. Other smaller-scale equipment is suggested in this chapter.

II. GRINDING

In most tropical countries, the soil is likely to be dry when dug, or it will dry out soon after digging. Even if it is wet, the best way to reduce it to a suitably fine size requires that it is first dried in the sun. In case of rainy weather, drying should be carried out in an open-sided shed. The dry soil may then be ground up.

An important principle to bear in mind in selecting a crushing or grinding method is the need to remove the material from the crushing zone as soon as it has been reduced to the required size. Thus, it is possible to handle soil which is still slightly damp. Furthermore, efforts will not be wasted in merely re-compressing the fine material into hard lumps. Some of the equipment for grinding and crushing suggested for small-scale production is described below.

II.1 Simple hand tools

The simplest device to break down lumps of soil is a punner (figure III.1). Basically, this is a hand-operated device comprising a flat-bottomed iron or hardwood weight attached to the end of a 1.5 m pole. The soil is spread out on a hard surface and the punner is raised and dropped repeatedly on the soil. Using a punner in not only hard work but also has one significant disadvantage: it recompacts the broken-down material which is then difficult to mix and process into blocks.

A useful multi-purpose hand tool, partly based on the punner, is the hammer-hoe (figure III.2). It may be used for two separate operations: the wooden mallet head can be used to break the soil and the metallic hoe blade can be used to move or mix soil. It has a slight advantage over the punner, in that broken soil can be moved more readily from the crushing area.

II.2 Pendulum crusher

The pendulum crusher is a labour-intensive crushing machine which has been developed in the United Kingdom. It is commercially available from licensed manufacturers in several countries. It is fairly suitable for small-scale production of stabilised soil blocks. Figure III.3 illustrates the crusher in operation. The pendulum crusher can be easily unbolted and transported on a small truck to another site. It works on the pendulum

Figure III.1
Punner for breaking down soil

Figure III.2
Hammer-hoe for breaking down soil

Figure III.3

One-man pendulum crusher in operation

principle. It can be fed and operated by a single worker, if necessary. The soil (from a wheelbarrow or heap on the ground), which is placed in a feed hopper at the top of the pendulum, comes into contact with a static grinding head and a curved moving grinding head. The latter is attached to the top of the heavy pendulum which is kept swinging by one or two people. The moving head is studded with protuding bolt heads which entrap and crush the soil as the head rotates in a downwards direction. The crushed soil drops through the small space between the fixed plate and the moving head. This space is adjustable and the moving head can be correctly aligned by moving the pivot bearings on the main frame. The ground soil falls by gravity on a built-in sieve of any desired mesh size. The sieved material is collected in a tray beneath the screen and runs down into a bin. Oversize material is collected as it runs off the top surface of the screen, and returned to the hoppers for further crushing. On the upwards return move, any remaining soil is cleared from the grinding surfaces prior to the next downward swing, so that a slight dampness of the soil is not a great problem.

The amount of crushed material passed through the screen can be increased by laying sacking or cloth on the screen to prevent particles from merely bouncing down and avoiding the holes. The cloth also reduces the amount of dust emanating from the machine.

Each time the container is full of finely ground soil, the operators would be well advised to change tasks in regular rotation as follows: 1. feeding soil; 2. pendulum handle (right side); 3. pendulum handle (left side); 4. attending discharge and resting; then back to feeding again.

An important part of the machine is the box beneath the moving head, filled with sand, gravel or any other available material in order to make it heavier. This weight, swinging as a pendulum, provides sufficient momentum to crush the harder soil particles without stopping the movement. If a virtually uncrushable piece of stone or debris is caught between the moving head and the fixed plate, the machine will not be over-stressed since the hard material will act as fulcrum for the pendulum, and the pivot will ride up the elongated bearing surface. As soon as the stone or debris has dropped through on the screen (and has been deposited with over-size material), the pivot resumes its normal position at the bottom of the bearing box.

Figure III.4 shows details of the components of the crusher, including the wire mesh safety screens and the cover for the pendulum box, complete with fastener. The moving head has a row of bent spikes above the bolt heads. These are useful in forcing into the crushing zone large pieces of soil which might otherwise merely ride on top of the moving head.

According to the material requirements of the block making process selected and the quality of product required, the particle size of the output can be chosen by adjusting the position of the pivot bearings and selecting an appropriate size sieving screen. Clayey soils with up to 18 per cent moisture content can be crushed satisfactorily in the machine.

Occasional greasing of the bearing is the only maintenance required, though the machine should be periodically inspected for wear, and parts replaced if necessary. This is likely to happen with the bolt heads in the moving head. All nuts, including those holding the replaceable bolts in the moving head should be checked for tightness, before first use and at regular intervals.

A larger machine, requiring a four-man team, is also available. It was the forerunner of the one-man machine.

Pendulum crushers, which have been operated in several countries, can be manufactured from readily available steel sections. Even the curved moving head can be made from approximately a dozen lengths of angle iron. Nevertheless, entrepreneurs should first refer to the innovators for precise details before adopting the method. Ready-made machines or sub-assemblies can also be purchased from the manufacturer (see Appendix III).

II.3 <u>Other hand-powered methods</u>

Metal rollers have been tested, but they do not crush effectively, especially if they are of small diameter. In this latter case, the rollers do not nip the soil particles easily but allow them to roll on top of the rollers without crushing. Furthermore, the rollers quickly become clogged if the soil is slightly damp.

A rotating metal drum, with part of the side replaced with a wire sieve, has been developed on an experimental basis. More testing is needed before it can be recommended.

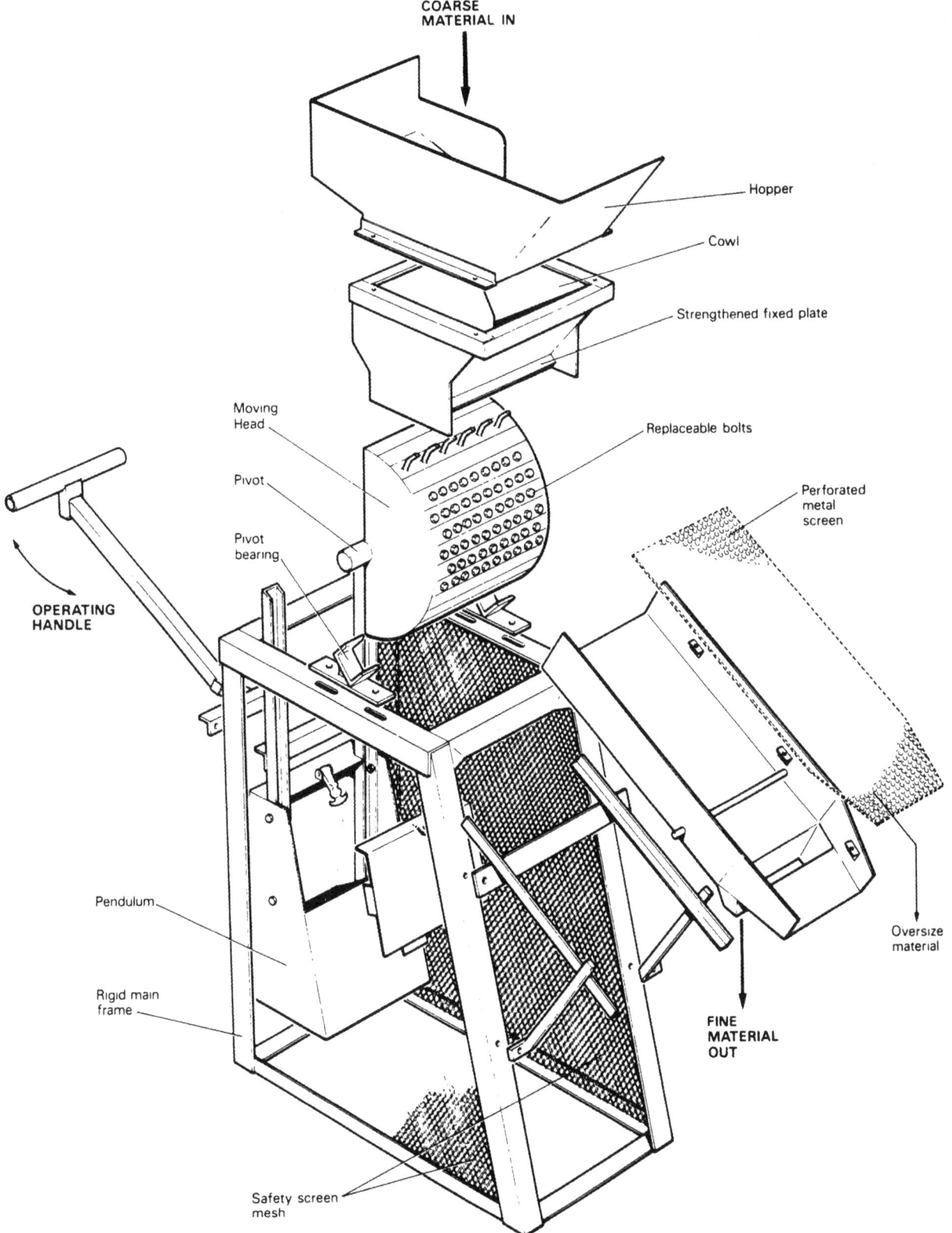

Figure III.4

Exploded view of a pendulum crusher

III. SIEVING

The material produced by crushing contains various sizes of material, from very fine dust up to pieces which are still too large for use in blockmaking. The oversize material must be removed by sieving, either through a built-in sieve, as with the pendulum crusher, or as a separate operation.

The simplest sieving device is a wire mesh screen, nailed to a supporting wooden frame and inclined at approximately 45° to the ground (figure III.5). The crushed material is thrown against the screen, the fine material passing through and the coarse, oversize material running down the front. Alternatively, the screen can be suspended horizontally from a tree or over a pit. This latter method is suitable in the only case where most material can pass through; otherwise too much coarse material is collected, and the screen becomes blocked and requires frequent emptying.

Devices such as the pendulum crusher, with a built-in screen, obviate further sieving and handling between operations.

IV. PROPORTIONING

Before starting production, tests should be made (see Chapter II) to determine the exact proportions of soil stabiliser and water for the production of good quality blocks. These proportions of materials and water will then be used in the production process. In order to ensure homogeneity of the blocks produced, the weight or volume of each material used in block making should be measured at the same physical state for subsequent batches of blocks. For example, the volume of soil or stabiliser should be measured in the dry or slightly damp state.

Once the exact proportions of each material have been determined, it is advisable to build a gauge box for each component (see figure III.6). The dimensions of each gauge box should be such that their content, when full (the material levels with the top edge of the box), should be equivalent to the fraction which should be mixed with other materials measured in other gauge boxes. Alternatively, a single gauge box may be used for all materials. In this case, the amount of material for the production of a given batch of blocks may be measured by filling and emptying the gauge box a number of times

Figure III.5
Simple screen to separate
fine material from coarse material

Figure III.6
Gauge box for measuring
quantity of materials

for each separate material. For example, a batch of blocks may require 10 gauge boxes of soil for one gauge box of stabiliser. Water may be measured in a pail or small tank.

It is advisable to mix a sufficient quantity of materials for the operation of the block making press (see next chapter) over approximately one hour. Thus, the volume of mixed material will depend on the hourly output of the press.

V. MIXING

It is most important that mixing be as thorough as possible in order to ensure the production of good quality, homogeneous blocks. Thoroughness of mixing is difficult to measure, though uniform colour of the mix may be a useful indicator when white lime is used as a stabiliser.

Dry components should be mixed first, then water added, and mixing continued until a homogenous mass is obtained. Mixing can be carried out by hand on a hard surface (concrete if possible), with spades, hoes, or shovels.

The necessary quantity of water must not be added all at once or to one part of the dry mix only. It is much better to add a little water at a time, sprinkled over the top of the mix, from a watering can with a rose spray on the nozzle. The dampened mix should be turned over several times with a spade or other suitable tool. A little more water may then be sprinkled on, and the whole mixture turned over again. This process should be repeated until all the water has been mixed in.

If lime is used as a stabiliser, it is advisable to let the mix stand for a short while before moulding starts to allow better moistening of soil particles with water. However, if cement is used as a stabiliser, it is advisable to use the mix as soon as possible, because cement starts to hydrate immediately after it is wetted and delays will result in the production of weaker blocks. This explains the earlier recommendation that the quantity of mix should not exceed what is needed for one hour's operation. Even so, the blocks produced at the end of an hour may be considerably weaker than those produced immediately after mixing.

A concrete mixer, even if available, will not be useful for mixing the wet soil, since the latter will tend to stick on the inside of the rotating drum. If machinery is to be used for mixing, it should have paddles or blades which move separately from the container. However, field experience shows that hand-mixing methods are often more satisfactory, more efficient and cheaper than mechanical mixing, and are less likely to produce small balls of soil which would be troublesome at the block forming stage.

VI. PRODUCTIVITY OF LABOUR AND EQUIPMENT

The rate at which a soil can be prepared depends upon its nature and the maximum size of grain acceptable after crushing. If the soil is fairly dry, lumpy and moderately hard, a team of four men, equipped with punners, can crush two tonnes of soil per day.

The one-man pendulum crusher may process 1.5 tonnes of soil per day under favourable conditions. The larger size pendulum crusher will produce approximately three times this quantity, but will require four operators (i.e. 4.5 tonnes per day).

Estimates of the productivity of the above soil crushing methods are provided in table III.1

Table III.1
Productivity of soil crushing methods

Method of crushing	Rate of production (man hours per tonne)
Punner	16
One-man pendulum crusher	5
Large pendulum crusher	7

VII. QUANTITY OF MATERIALS REQUIRED

Stabilised soil blocks are usually larger than traditional burnt bricks. A typical block size is 290 x 140 x 90 mm. Its production will require 7.5 to 8 kg of material. The exact quantity of stabiliser necessary must be determined for any particular project, by means of the tests described in Chapter II. The fraction of lime or cement usually varies between 5 and 8 per cent. Similarly, the optimum moisture content for any particular soil must be determined experimentally. The moisture level will vary widely with the nature of the soil. An approximate estimate of 15 per cent by weight is often assumed.

The quantities of materials required for a typical block press producing 300 blocks per day are shown in table III.2 below.

Table III.2

Approximate quantities of materials required for producing 300 blocks per day

Material	Quantity required per day	
	8 per cent hydrated lime	5 per cent ordinary Portland cement
Soil	1.9 tonnes	1.95 tonnes
Stabiliser	150 kg	95 kg
Water	300 litres	300 litres
Total (after mixing)	2,350 kg	2.345 kg

In practice, the quantities of soil, stabiliser and water required will vary from the above estimates, depending upon the type and properties of the soil. A single-storey house covering an area of 50 m^2 will require approximately 3,000 blocks. The estimated quantities of soil, stabiliser and water required for the building of such a house are provided in table III.3. In this example, the blocks for the house could be produced in 10 days.

Table III.3

Approximate quantities of materials for a single storey house (50 m^2 plinth)

Material	Quantity required per house	
	8 per cent hydrated lime	5 per cent ordinary Portland cement
Soil	19 tonnes	19.5 tonnes
Stabiliser	1.5 tonnes	0.95 tonnes
Water	3000 litres	3000 litres

CHAPTER IV

FORMING

I. <u>BUILDING STANDARDS AND BLOCKS</u>

Several factors should be considered before starting a stabilised soil block operation. These include: the type of stabiliser to be used; whether the soil is suitable for stabilisation; whether the formed block will meet local building standards and whether stabilised soil blocks will be strong enough to be used as load-bearing elements.

One of the aims of this memorandum is to make the reader aware of the problems associated with the use of soil in the construction industry, especially in developing countries.

In the majority of developing countries, building standards are not yet developed or applied, especially in the field of soil construction. A number of current soil construction techniques are inefficient and wasteful of resources. The quality of the building materials produced can also be improved.

In view of the above inefficiencies, the International Union of Testing and Research Laboratories for Materials and Structures (RILEM), formed in 1983, a technical working committee for "laterite-based materials". The objective of this committee was to produce an international draft building standard covering the use of stabilised soil building blocks. However, it is becoming increasingly clear that it is difficult to propose one set of building standards to meet all requirements throughout the world. For example, a minimum wet strength of about 1.4 MN/m^2 has been recommended by a number of building authorities, while the soil brick specification of the state of New Mexico (United States) states that the average compressive strength of rammed earth soil bricks should be 2.04 MN/m^2, and that only one out of five blocks may have a compressive strength of not less than 1.63 MN/m^2.

There is, in general, a wide variation of acceptable standards which reflect, to some extent, local weather conditions. Blocks with wet compressive strengths of 2.8 MN/m^2 or higher (i.e. minimum requirement for fired bricks and concrete blocks - see Chapter I, section III.1) should be suitable for one and two-storey buildings. Furthermore, they would probably not require external protection against the weather. For one-storey buildings, blocks with a compressive strength of 2.04 MN/m^2 would probably be strong enough, but where rainfall is high, an external protective coating may be required. Since the wet strength of a stabilised soil wall may be less than two-thirds of its dry strength, all compressive strength tests should be performed on samples which have been soaked in water for a minimum of 24 hours after the appropriate curing period.

The final wet compressive strength of a soil block depends not only on the type of soil but also on the type and quantity of stabiliser that has been used, the forming pressure used to mould the block, and the subsequent curing conditions.

As stated earlier, the wet compressive strength of a stabilised soil building block is determined after the block has been totally immersed in water for a period of 24 hours. If the block is weighed before and after immersion, a moisture absorption figure can be determined.[1] If this figure is greater than 20 per cent, the resulting external wall built with this type of block may need an external rendering to improve its long term durability.

II. COMPRESSIVE STRENGTH : DENSITY AND MOULDING PRESSURE RELATIONSHIPS

Before discussing the principles involved in forming a stabilised soil block it is useful to analyse the relationships between the following variables: (i) the compressive strength of a block; (ii) its density and (iii) the moulding pressure used to make a block. These relationships have been investigated in a study on stabilised soil construction published by the United Nations in 1958[2] The study describes tests performed on two different types of soil from Burma, each stabilised with 5 per cent cement.

---

[1]This figure is equal to the percentage increase of the weight of the dry block after immersion in water for 24 hours (see Chapter V).

[2]The study is described in Fitzmaurice, 1958.

The first relationship cited in the study is shown in figure IV.1 where the dry density is plotted against the dry compressive strength of the block.[1] It can be seen that the relationship between dry strength and density is almost linear. It may be stated that <u>the strength or durability of a block increases as the dry density increases</u>.

The second relationship cited in the study is shown in figure IV.2 where the dry density is plotted against the compaction or moulding pressure for a series of tests at various moisture contents. This set of results is important because various machines and various methods of compaction will yield different results in terms of compressive strength. The main conclusion derived from the tests is that <u>dry density increases as the compacting or moulding pressure increases. Dry density is also dependent upon the moisture content of a mix.</u> It can be seen that, for a given compaction pressure, the dry density generally increases as the moisture content decreases.

The third relationship cited in the study is shown in figure IV.3, where the dry compressive strength of a block is plotted against the percentage moisture content of the mix. One set of blocks was made in a hand-operated toggle press which was believed to have a compacting or moulding pressure of about 4 MN/m^2; the second set of blocks was made in a hydraulic power-driven press, exerting a compacting pressure of about 7MN/m^2. Results from the tests clearly indicate that the higher the compacting pressure the higher the dry compressive strength. They also show that <u>the optimum moisture content (OMC) decreases with an increase of the compaction pressure</u> (refer to points A and B in figure IV.3).

The importance of the moisture content at the time of testing has been emphasised by a number of authors[2]. Stabilised soil, in common with other porous building materials, is very sensitive to moisture content. The wet compressive strength is always considerably lower than the dry strength. It would be unwise to assume that a wall or pier will never get wet over the entire life of a building. For example, tests carried out on the Burmese soil

---

[1] The specimen used for the study are soil cylinders with a diameter of 76 mm and a length of 80 mm. The cylinders are crushed dry after a curing period of two months.

[2] See Fiztmaurice, 1958.

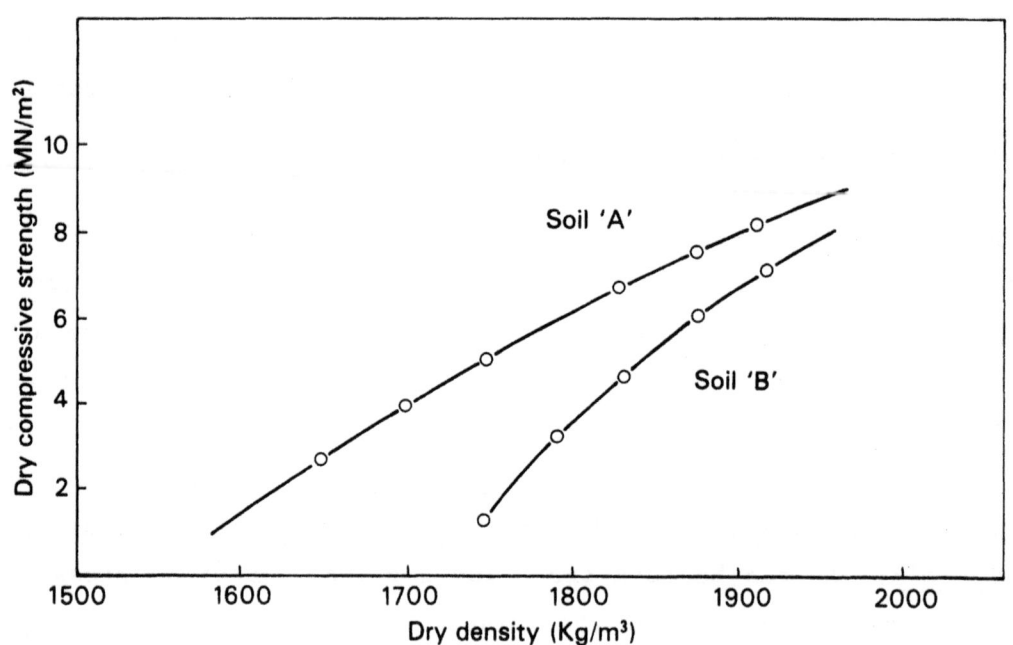

Figure IV.1

Dry compressive strength-dry density curve

(Source: Fitzmaurice, 1958)

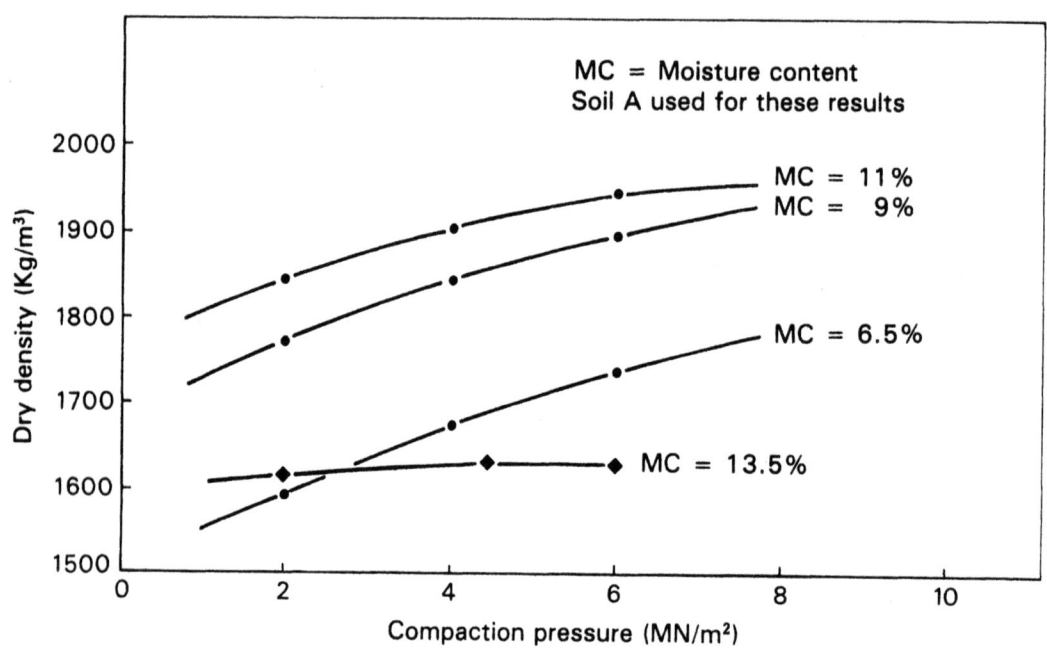

Figure IV.2

Relation between dry density, compaction pressure and moisture content

(Source: Fitzmaurice, 1958)

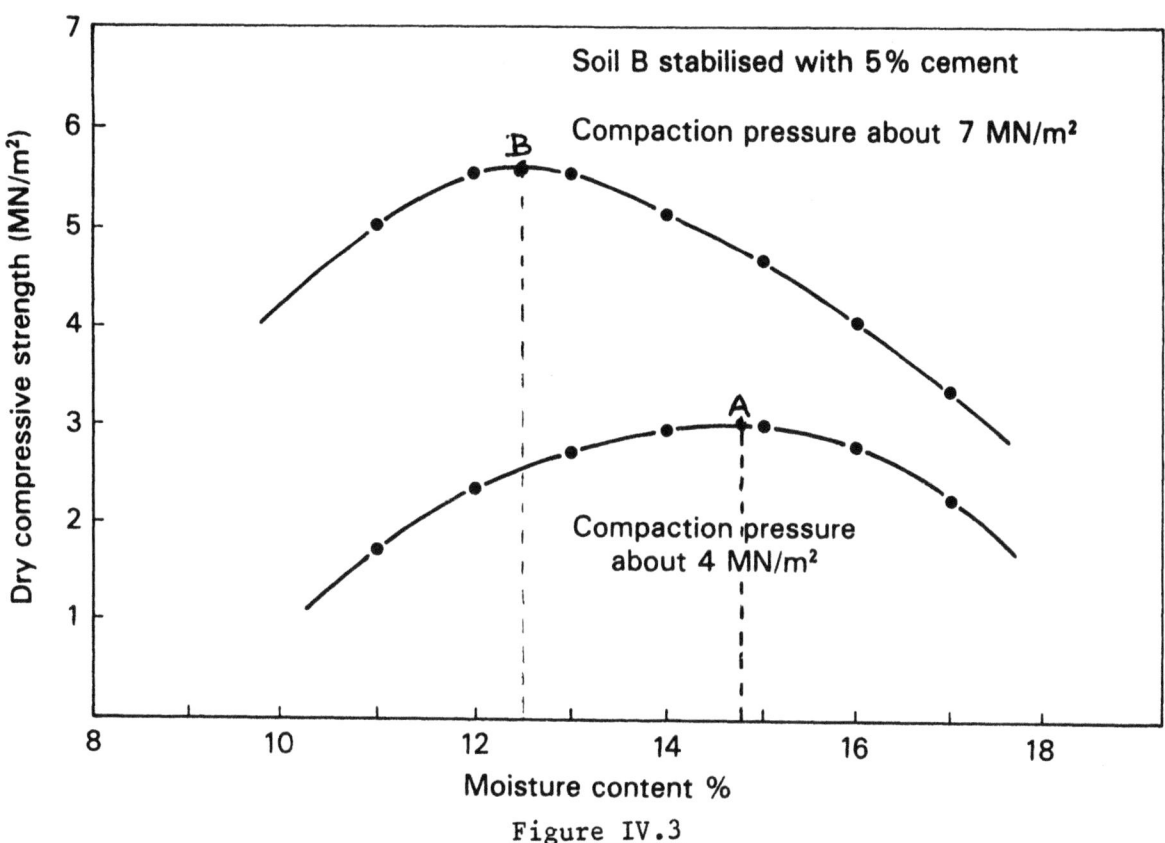

Figure IV.3

Relation between dry compressive strength, moisture content and compaction pressure

(Source: Fitzmaurice, 1958)

(stabilised with 5 per cent cement) show that the wet compressive strength was in the range of 40 to 50 per cent of the air dry compressive strength. Consequently, buildings should always be designed on the basis of the wet compressive strength.

III. BLOCK FORMING METHODS

There are two basic methods of forming a block:

- at constant pressure; and
- at constant volume.

These two methods are briefly described below.

Constant pressure method

When using the constant pressure method to form or compact a soil mix in a mould, the mix is subjected to a uniform pressure which reduces the air voids in the material by moving the soil particles together. This internal soil particle movement lasts until the soil mix develops an internal pressure equal to that of the externally applied pressure, at which time no further compaction movement takes place.

The fixed external pressure, the moisture content of the soil mix and the initial quantity of material deposited into the mould determine the thickness of a block. The production of uniform blocks therefore requires that both the quantity and moisture content of the soil remain constant at all times.

Constant volume

When using the constant volume method, the external compacting pressure varies so that blocks of uniform thickness are produced even though the quantity of soil may vary. However, variation in the quantity of soil results in a variation in the density of blocks produced. Since density affects durability, a wall constructed with blocks of variable density will, in time, suffer from uneven erosion. Therefore, in both the constant pressure or constant volume methods, the weight or volume of a soil mix fed into a mould, as well as the moisture content of the mix, should be kept uniform. Frequent checks should be made to ensure the production of blocks of uniform thickness and density. A device that operates both at constant pressure and at constant volume produces more uniform blocks and is ideal for the production of stabilised soil blocks.

IV. SOIL TESTING PRIOR TO PRODUCTION

Chapter II described various soil testing procedures for determining whether a soil is suitable for block making and the type and amount of stabiliser which should be added. These tests are performed first, since positive results are a pre-requisite for setting up a production unit. Once production starts, the soil mix must be checked for each batch of blocks to determine its moisture content: the latter should be as close as possible to the optimum moisture content (OMC). For this purpose, two simple ad-hoc site spot checks can be performed. These are described below.

(i) Pick up a handful of soil mix and squeeze it in the hand; the mix should "ball" together and, when the hand is opened, the fingers should be reasonably dry and clean;

(ii) Drop the "balled" sample onto a hard surface from a height of about one metre:

- if the sample completely shatters on impact, this indicates that it is not sufficiently moist;

- if the sample "squashes" into a flattened ball or disc on impact with the hard surface, this indicates too high a moisture content;

- if the sample breaks into four or five major lumps, this indicates that the moisture content of the soil mix is close to the optimum moisture content (OMC).

The soil mixture can then be used for block making. However, to produce blocks of uniform size and density, special precautions must be taken to fill the mould with the same quantity of mix at each pressing. It is thus recommended to pre-weigh each mix. If this is not practical, a small wooden box or tin may be used to ensure that the same volume of mixed soil is used.

A few experimental pressings must be conducted before the correct amount of mixed soil is determined. It is also essential to consult the block machine manufactuer's operational manual in order to ensure that the block making machine is properly used.

To facilitate demoulding of the blocks and to ensure good clean surfaces and arrisses, it is advisable to moisten the internal faces of the machine's mould with a mould release agent (usually a form of oil), which can be applied with either a rag, brush or spray.

For low pressure block making machines (employing up to 2 MN/m^2 compaction pressure), a mould release agent can take the form of a liquid mud mix. The latter may be simply made by adding a large amount of water to part of a soil mix. For higher compaction pressure machines (operating up to about 15 MN/m^2), waste engine oil has often proved satisfactory. Several other mould release agents can be employed (e.g. diesel, kerosene, coconut oil or even liquid detergent). However, used engine oil should be both cheap and easily available.

Experience has shown that the mould should be oiled about every fourth pressing. Approximately 250 blocks can be produced from one litre of used engine oil. The quantity of mould release agent required will depend on the absorption characteristics of the soil.

The application of a mould release agent to the walls of the mould will ensure easier demoulding and produce stabilised soil blocks which better withstand weathering in the field.

If special blocks are to be made (e.g. hollow, grooved or frogged blocks), simple wooden forms should be used. These forms should be coated with a mould release agent at each pressing.

The mould box should be evenly filled and the corners well packed with a pre-determined quantity of soil mix. To obtain a good density block, it is advisable to compress the soil mix in the mould lightly by hand.

V. BLOCK SIZES

The overall dimensions of a block should suit the appropriate system of modular co-ordination in order to reduce the need for excessive cutting or the provision of special-sized infill blocks. The length and width are usually appreciably greater than those of the standard size brick. There are two main reasons for this larger block size: to increase the productivity of masons in wall construction and to reduce the volume of mortar used to cover joints.

Adobe (non-stabilised soil blocks) are normally square in shape and vary in size between 300 x 300 mm and 400 x 400 mm. These dimensions are usually required in view of the relatively low density and strength of these blocks. The relatively large block area results in a lower compressive stress to carry the vertical and lateral loads imposed by the total building weight.

Most countries currently use concrete blocks 400 mm long and 200 mm high, with varying thicknesses up to a maximum of 200 mm. These dimensions are not feasible for stabilised soil blocks because the production of good-quality blocks of this size requires relatively high compacting forces. It is therefore necessary to adopt smaller overall dimensions. It is traditional to lay concrete blocks with 10 mm thick mortar joints. If this is acceptable practice for wall construction, a stabilised soil building block 290 mm long, 140 mm thick and 90 to 100 mm high would be acceptable. With this suggested block size, the following standards should be met:

- with a mortar joint of 10 mm, a module size of 300 mm could be used;
- a double skin wall thickness of 290 mm would be possible;
- if a minimum wet compressive strength of 2.8 MN/m^2 is achieved, a single skin wall thickness of 140 mm would be sufficient to carry the vertical and lateral loads in a single storey building (and probably two-storey buildings), provided the foundations are sufficient;
- good durable features should be achieved without the need for costly external protective renderings to resist weathering problems;
- with a density of about 2,000 kg/m^3, an individual block will have a dry weight of about 7 kg which is easy for the mason to handle; and
- a wall thickness of 140 mm with a density of 2,000 kg/m^3 should provide adequate thermal insulation even when external wall temperatures fluctuate widely. Furthermore, high thermal capacity will be obtained which should help reduce temperature variations inside a building.

VI. PROPOSED TECHNICAL STANDARDS FOR COMPRESSED SOIL BLOCKS

CRATERRE[1], the international centre for research and the application of earth construction, recently proposed technical standards for lime stabilised compressed soil blocks. These standards are derived from a study of soil block making machines. They are reproduced in this section with minor alterations suggested by the authors of this memorandum.

Block dimensions

The study related to blocks which were parallelepipeds with the following maximum dimensions:

Length: 400 mm (exceptionally 500 mm);
Width : 200 mm (exceptionally 300 mm);
Height: 200 mm.

---

[1] The acronym CRATERRE stands for "Centre de recherche et d'application-terre" (see Appendix II).

A height of more than 200 mm would make an individual block too heavy for a mason to handle efficiently.

Currently manufactured block types have the following nominal dimensions:

Length: 295 mm
Width : 140 mm
Height: 88 mm

Dimensional tolerances:

Length : + 1 mm; − 3 mm
Width : + 1 mm; − 2 mm
Height : + 2 mm; − 1 mm
Surface smoothness sides: + 1 mm; − 1 mm
Compression surfaces: + 3 mm; − 1 mm

Edge smoothness

The maximum sweep for edge smoothness is 2 mm. A roughness is tolerated as long as it is due to demoulding and manipulation. It may be noted that roughness of upper and lower block faces improve mortar joint adhesion as well as the shear resistance of a wall.

Caverns, holes, alveoles

Caverns, holes and alveoles are tolerated on the same terms as smoothness. The following standards are suggested: defects covering less than 1 per cent of exposed surface and less than 15 per cent of non-exposed surface.

Specific density

The suggested specific densities of blocks are shown below:

dry blocks :   − Minimum   : 1,700 kg/m^3
                          − Recommended: 2,000 kg/m^3
wet (freshly moulded)
blocks   − Minimum   : 1,870 kg/m^3
                          − Recommended: 2,200 kg/m^3
nominal volume of blocks: : 3.634 litres.

Skewness of surfaces

A standard skewness of surfaces is recommended by the study. The exterior faces may be slightly oblique if prescriptions of dimensions, tolerances and forms are respected. The interior surfaces of hollow blocks should preferably be oblique. This is most desirable because it allows for easy demoulding immediately after compaction. The interior spaces of hollow or alveolar blocks may not have sharp corners.

Rugosity of exterior faces

The exterior face of blocks to be coated with mortar or renderings should preferably be rugous, while those which do not receive a coating must be smooth.

Clefts - scaling

These are not tolerated on any surface.

Gaps, cracks, crevices

Micro-cracks are tolerated on all surfaces; macro-cracks are tolerated only on non-exposed surfaces. The width and depth of these cracks may not exceed 1 mm whilst the length may not exceed 10 mm. The total number of cracks may not exceed the average value of one per 100 mm rib length.

Chipped edges

The width and depth of chipped edges may not exceed 10 mm.

Wall thickness of alveolar or hollow blocks

For all faces, the minimum thickness of solid material surrounding the alveoles or hollow blocks should be as follows:

- 35 mm for low pressure blocks (20 da N/cm^2 or 2 MN/m^2); and
- 20 mm for high pressure blocks (100 da N/cm^2 or 10 MN/m^2).

Minimal proportion of the load-bearing surface to the nominal surface of hollow blocks

The minimal proportion of the load-bearing surface to the nominal surface of hollow blocks varies with the compaction pressure used for manufacturing the blocks. It is superior or equal to 0.6 for low pressure blocks and superior or equal to 0.4 for high pressure blocks.

Scratches

The following standards are suggested for scratches:

- maximum depth: 10 mm;
- maximum width: 15 mm;
- maximum area of scratches on surface: 100 mm^2; and
- minimal distance between the edge and a deep scratch: 35 mm.

Special blocks

Special blocks may be produced for specific purposes. Some of these are briefly described below.

Blocks with <u>differential stabilisation</u>: these have one or more surfaces or parts which contain more stabiliser than the rest of the block.

Blocks with <u>built-in facing tile</u>: these blocks have one or more surfaces decorated with a special facing tile.

Blocks with <u>treated surface</u>: these blocks have one or more surfaces especially covered with graphic elements or decorative elements treated with a chemical.

Resistance (compressive strength)

The following compressive strengths of stabilised soil blocks are adopted by a large number of countries:

- the dry compressive resistance after 28 days must be equal or superior to 2.1 MN/m^2;

- the wet compressive resistance after 28 days (saturated humidification) must be equal or superior to 1 MN/m^2.

The wet compressive resistance quoted above may be suitable for a dry arid zone but an external rendering coat of material would certainly be needed for weather protection. If it is possible to manufacture stabilised soil building blocks with a wet compessive strenght of 2.8 MN/m^2, an external rendering application is not required.

When properly conducted tests have shown that wet compressive strengths approaching 2.8 MN/m^2 can be obtained, it is appropriate to design higher building stresses and therefore accept the value of 2.8 MN/m^2 as standard.

VII. SOIL BLOCK MAKING MACHINES

Although soil has been used as a building material for a very long time, variable climatic conditions have prevented a general adoption of this material, especially in temperate climates.

The production of acceptable quality stabilised soil blocks requires that soil mixes be compacted in order to reduce the air voids within the material and thus improve the strength of the block.

There are two basic methods of moulding a soil block:
- use of soil block presses; or
- casting a mud mix in forms or moulds by hand, using a tamping method.

The adobe block is usually produced with the second method, whereby water and a sandy clay soil are mixed into a mud consistency and formed into blocks. Chopped straw is often added to the mix to reinforce and minimise the drying out shrinkage cracks which will otherwise occur. Adobe block manufacture is illustrated in figure IV.4. The mix is thrown into a simple, open-topped wooden or steel mould form and tamped or pressed by hand to fill the mould space completely. The form is then removed and the operation repeated. After demoulding, the formed block is allowed to dry in the sun.

Sophisticated concrete block making machines exerting compacting pressures of up to 16 MN/m^2 have been developed. They produce either a single or several blocks in a single operation. In the latter case, they are

Figure IV.4
Adobe block manufacture

called egg-laying machines. These machines, usually expensive, use both direct pressure and vibration and are not suitable for the production of stabilised soil blocks: concrete mixes have a moisture content of about 40 per cent, while stabilised soil mixes have a moisture content of about 15 per cent. Different machines have therefore been developed for the production of stabilised soil blocks. Some of them are described below.

VII.I The CINVA-Ram press

In the early 1950s, an engineer[1] employed by the Inter-American Housing and Planning Centre (CINVA) in Bogota, Colombia, developed a constant volume soil block making machine which has since been known as CINVA-Ram. This machine is illustrated in figure IV.5.

---

[1] Paul Ramirez from Chile.

Figure IV.5
The CINVA-Ram block making machine

The CINVA-Ram block press consists of a mould box in which a slightly moist soil mix is compressed by a hand-operated toggle lever and piston system. This machine has a tare weight of about 60 kg and employs a maximum compacting pressure of about 2 MN/m^2. It could thus be classified as a portable tool for a "do-it-yourself" builder for constructing small houses, walls and farm buildings. The all-steel machine produces blocks 290 mm long, 140 mm wide and 90 mm thick.

This machine has been used extensively in developing countries for the production of stabilised soil building blocks, with mainly cement used as a stabiliser. The following points are worth bearing in mind regarding the use of the machine:

- the initial amount of soil put into the mould box should be closely controlled; and

- the press will not have a long life if it is mishandled on a building site.

The VITA publication, <u>Making building blocks with the CINVA-Ram press</u>[1] indicates the following advantages of this block making machine:

- stabilised soil blocks are easier to make than concrete blocks: they can be removed immediately from the press and stacked for curing without a pallet;
- the cost of building materials is greatly reduced, since most of the raw material is locally available;
- transport costs are reduced, since the machine is portable and the blocks are produced near to the construction site;
- if the quality of materials used is good, CINVA-Ram blocks can be superior to adobe and rammed earth;
- blocks are easily handled;
- blocks need no baking, since the curing process is completely natural; and

---

[1] See VITA, 1977.

- the press makes variations of the blocks for the various phases of construction.

VII.2 The CETA-Ram press

The CETA-Ram press was developed by engineers from the Engineering Faculty of the San Carlos University (Guatemala) and researchers from the Centre of Appropriate Technical Experimentation (CETA-Guatemala). It is a modified CINVA-Ram press which allows the production of stabilised soil blocks with vertical holes. These blocks may then be used with vertical steel reinforcements in walls designed to better withstand earthquakes.[1]

A CINVA-Ram press was modified to manufacture stabilised soil blocks 320 mm long, 152 mm wide and 110 mm thick, with two 60 mm diameter holes passing through the thickness of the blocks. This machine, named the CETA press, is illustrated in figure IV.6. It is composed of three main parts:

- a main frame with the upper part forming the walls of a mould; the latter is fitted with a cover that swivels through 90°;

- a movable mould base plate which acts as a piston within the mould body; and

- the toggle mechanism and hand-operating lever.

Prototype CETA-Ram presses have been used extensively on an experimental basis in the building of rural housing. Results from these experiments show that the use of stabilised soil hollow blocks in the building of walls which must be reinforced with steel bars has two main advantages: it speeds up the work and reduces the cost.

In Guatemala, the CETA-Ram press was used for the production of blocks made from one part of cement and eight parts of volcanic material of the pumice type available in large quantities in the country. Produced blocks had compressive strengths ranging from 2.89 MN/m^2 to 6.8 MN/m^2. It is not specified whether these strengths apply to wet or dry blocks.

---

[1] The CETA-Ram press was developed in 1976, soon after the earthquake which struck Guatemala. The technological innovation was therefore in response to a real and pressing need to build houses which can better withstand the devastating effects of earthquakes.

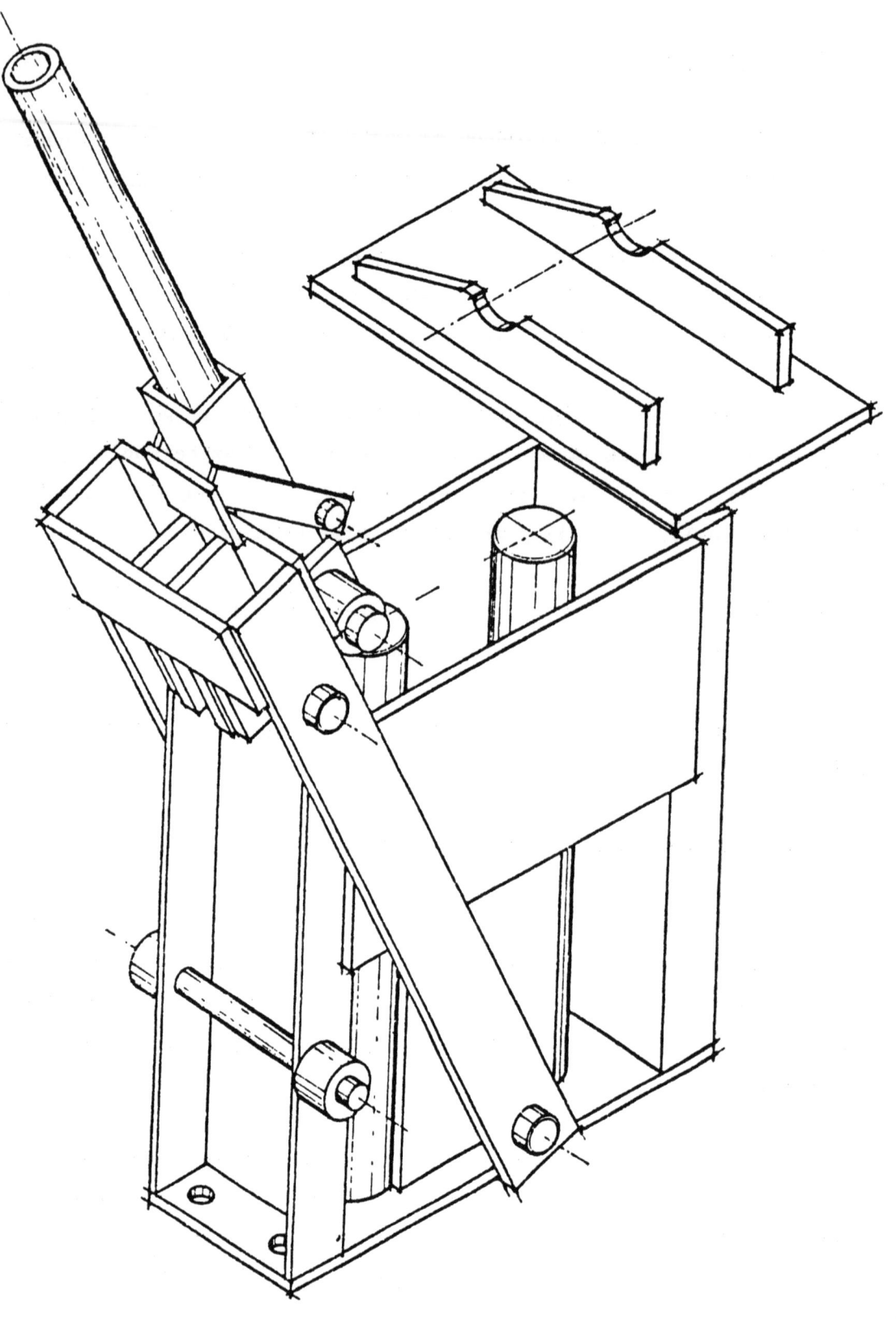

Figure IV.6

The CETA-Ram press

VII.3 Landcrete press/Presse Terstaram

The Landcrete block making machine was originally developed by Landsborough-Findlay Ltd. in the early 1950s. Two main models were introduced: a hand-operated toggle mechanism machine and a power-driven version. Both models are sturdy in construction and, according to the manufacturers, simple to operate. However, all references to this type of press are to be found in old literature. Several of the original Landcrete machines have been seen by the authors: in each case the machines were broken and were not operational.

The Landcrete press was partly redesigned and is now available from Belgian manufacturers under the name of "Presse Terstaram". It uses a compacting pressure of about 4 MN/m^2 and can produce various sizes of stabilised soil blocks from a 295 x 140 mm mould. It weighs about 350 kg. Figure IV.7 illustrates the Terstaram block-making machine. It shows two operators applying the main compacting force (of 20 tonnes) via a lever arrangement to compact a soil mix. The compacting pressure developed in the machine shown in figure IV.7 is 2.25 MN/m^2, a marginally greater pressure than that applied by the CINVA-Ram press.

VII.4 Tek-Block press

The Tek-Block press was developed by the University of Science and Technology of Kumasi (Ghana). This hand-operated press is illustrated in figure IV.8. It was supposed to replace the previously used Landcrete machine considered unsuitable for Ghanaian conditions.

The Tek-Block press was supposed to overcome the following deficiencies of the CINVA-Ram press:

- some of the materials used on the CINVA-Ram press were too thin in section and tended to deform after relatively short periods of use;

- the adjustable piston guides did not perform well and were poorly adjusted by the workers in the field;

- the top plate locking arrangement of the mould was too weak and could be automated; and

Figure IV.7

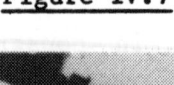

Presse Terstaram block making machine

Figure IV.8
The Tek-Block press

- the mould size (290 x 140 x 100 mm thickness) was rather small considering the labour involved. It could be made larger.

Consequently, the Tek-Block machine is made almost entirely of 12 mm steel plate. It cannot be adjusted on site and makes a block size 290 mm long, 215 mm wide and 140 mm thick. It uses the same toggle mechanism as that of the CINVA-Ram press but the main operating lever arm is 2.4 metres long and is made from timber. Thus, if the mould is overfilled, the timber lever arm would break before any damaging stresses would be incurred by the machine. The compacting pressure of the Tek-Block press is about 1.5 MN/m^2.

An additional major innovation concerns the covering lid of the mould; it is mounted on the upper handle socket assembly, and may thus be moved away from the mould with a movement of the main operating lever. The Tek-Block machine weighs about 90 kg. The first units of the Tek-Block machine tended to crack and some welds failed. These failures could be avoided with careful manufacturing.

Early site observations showed that a crew of five men and one Tek-Block machine could produce 150 to 175 blocks per day, if given proper incentives, whereas the manufacturers' handbook claims a daily output of 200 to 400 blocks.

A powered version of the Tek-block press was developed in the late 1970s. It proved too expensive and the project was terminated.

VII.5 Winget block making machine

The first Winget, shuttle mould, hydraulic block making machine was developed in 1948 and tested in the United Republic of Tanzania where good-quality, stabilised soil blocks 305 mm long, 150 mm wide and 100 mm thick were produced. The compaction pressure was 9.45 MN/m^2. The blocks produced had a satisfactory dry crushing strength of about 5.8 MN/m^2 after a period of 21 days. A medium-range, clay-content soil was used with a 2.5 per cent addition of cement. Despite these excellent results, it became obvious that profitable production necessitated an increase in the machine output and re-design of some of its parts. This resulted in the development of the Rotary Hydraulic Block Press machine (illustrated in figure IV.9) which is claimed to have a consistently high output rate of 140 blocks per hour. This machine has a tare weight of about 1,150 kg and uses a compaction pressure of 9.5 MN/m^2.

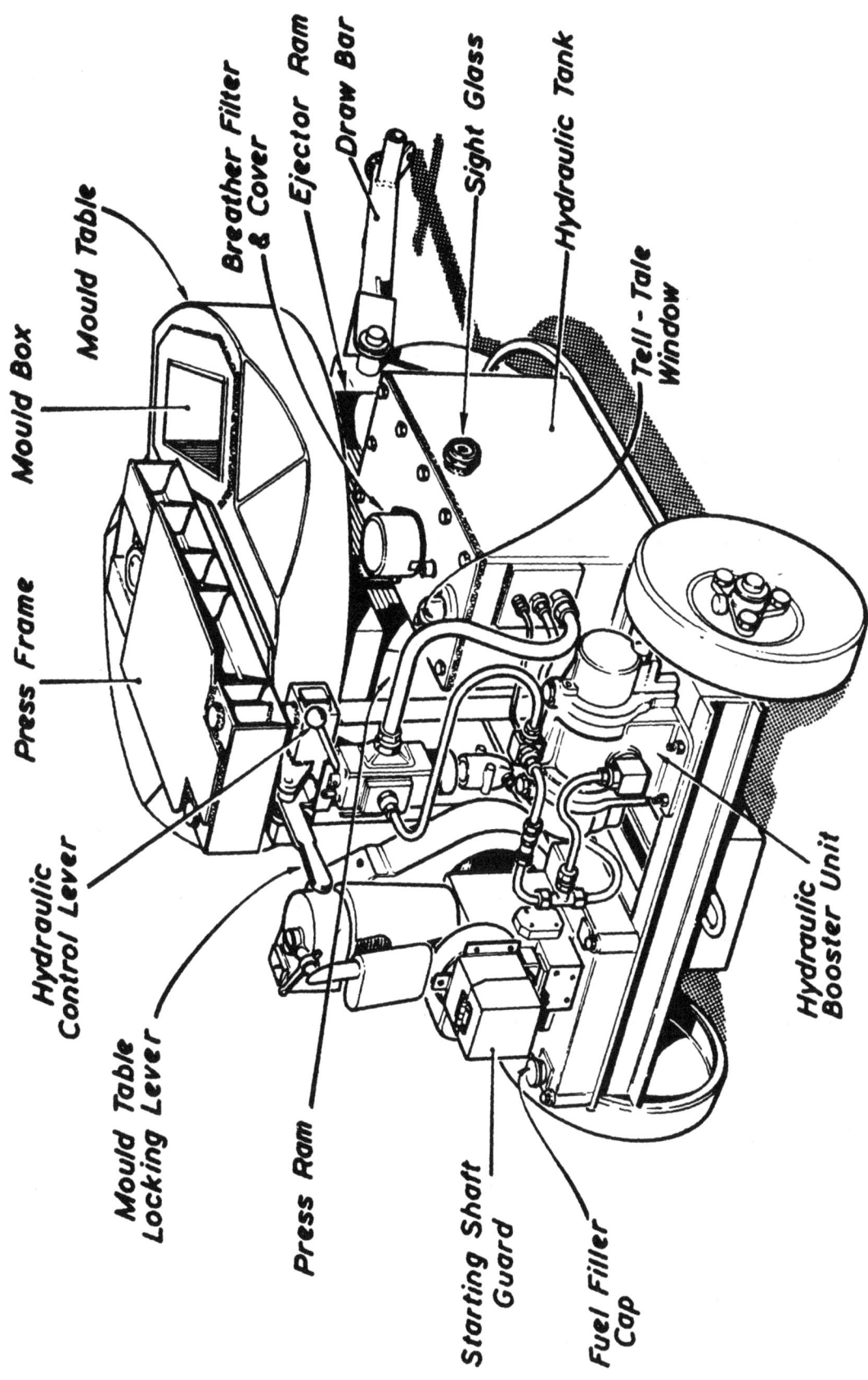

Figure IV.9

Winget rotary table mould machine

About 350 of these machines have been sold to some thirty developing countries. Owing to the high compaction pressure, the quality of blocks produced is very good and the production rate is three to four times greater than for a hand-operated machine. One disadvantage of this machine is the need to exert relatively high pressures which could damage the machine if it is not handled by skilled operators. In view of the high initial cost of this machine, demand decreased to such a level that it was decided to discontinue production in the early 1970s.

VII.6 Ellson Blockmaster Stabilised Soil Block Press

The Ellson soil block press was originally developed by Ellson Pty. Since 1978, it has been manufactured under license by Joshi Industries, Rajkot, Gujarat State, India. The latter firm renamed the machine the "Ellson Blockmaster Stabilised Soil Block Press" (illustrated in figure IV.10).

The Ellson Blockmaster machine is an all-steel welded assembly, manually operated, which can produce block sizes of either 290 x 190 x 90 mm thickness or 290 x 140 x 90 mm thickness. It has a tare weight of about 210 kg.

The lever is usually operated by two men who stand on the projecting inclined leg ready for the pull down stroke; these men must apply considerable effort in order to achieved a maximum compaction pressure of about 7 MN/m^2, although a leverage ratio of 500 to 1 is used. One significant feature of this machine is the height of the mould from the ground (about 850 cm). This height helps to reduce operator back-ache from bending down too low to remove freshly made soil blocks from the machine.

The manufacturer claims that a labour force of ten men is necessary to produce 750 blocks (290 mm x 190 mm x 90 mm) per day. This includes winning the soil; spreading it out for drying; sieving; mixing; filling the mould; pressing; and carrying away newly pressed blocks for stacking. Two of these machines have been seen in operation by one of the authors. In both cases, the daily output was in the range of 250 to 300 blocks. Each machine had to be rewelded at the lower end of the main operating lever on several occasions. This was due to the high stresses generated during the compaction stroke.

Depending on the nature of the soil and the stabiliser used, the manufacturer claims that well-stabilised dense blocks can be produced, and that the dry crushing strengths of these blocks vary between 4 and 12

Figure IV.10

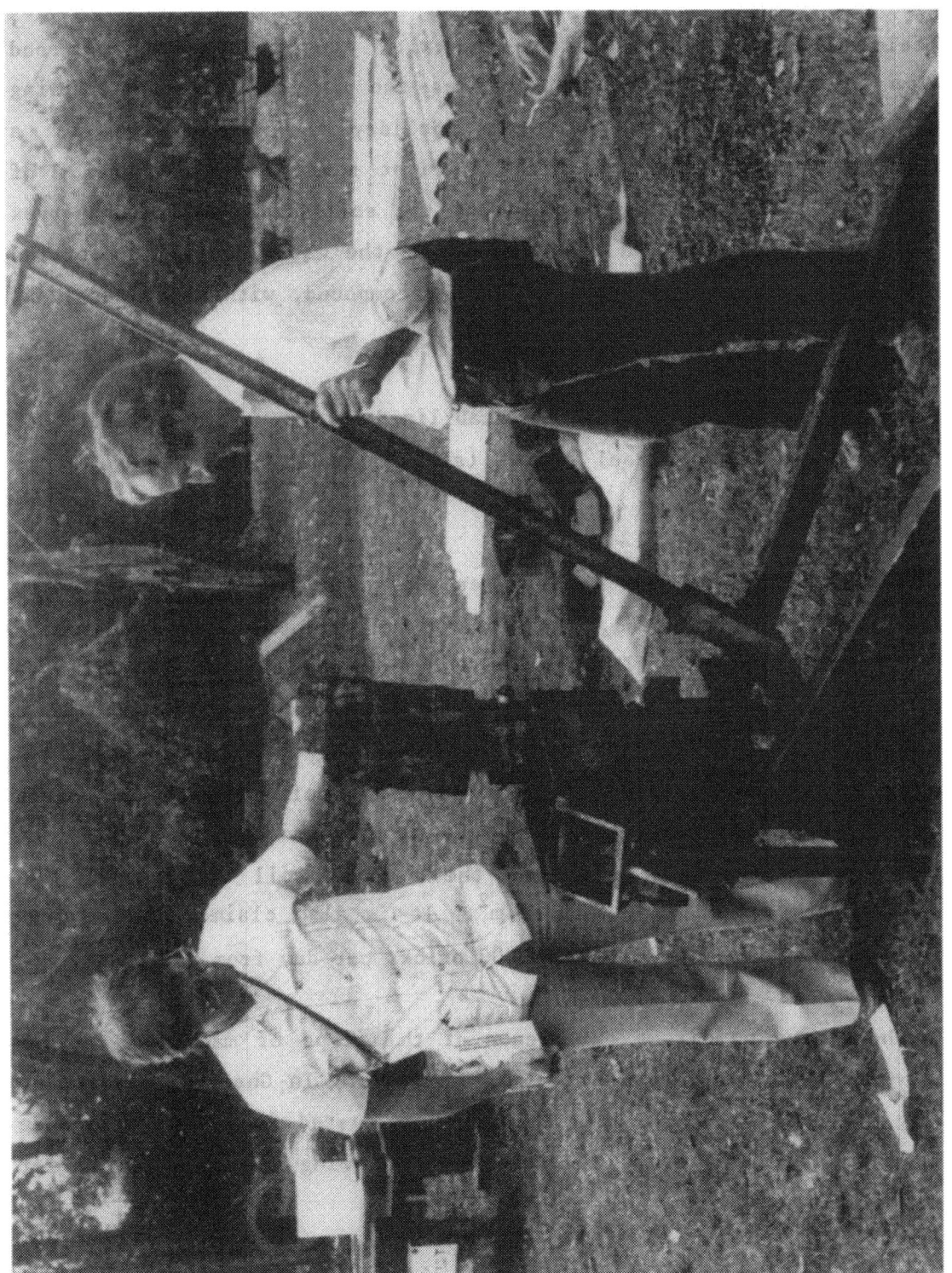

Ellson Blockmaster Stabilised Soil Block Press

MN/m^2. Moisture absorption is much lower than that of ordinary burnt clay stock bricks.

VII.7 Consolid AG

During the late 1970s, Consolid AG of Switzerland developed a new process of chemical soil stabilisation for use with cohesive soils on road construction projects. This process involves the use of three stabilising agents : Consolid 444, Conservex and Solidry. Consolid 444 is a silicone-copolymer resin solution which is first mixed with a quantity of water appropriate to the moisture content of the soil being used. Conservex is a type of bituminous emulsion used to enhance the waterproofing properties of Consolid 444. Solidry is a powdered polymer compound, with water resistant properties.

Consolid AG developed a mobile, stabilised block making plant "CLU 3000". Powered by a 13 hp diesel engine (see figure IV.11), it has a tare weight of about 1,600 kg.

This trailer plant comprises a diesel engine, paddle mixer and feed unit, four cavity rotary table press, soil mixer and the necessary hydraulic components used for pressing. The pressing of a brick is manually initiated by the operator: the mould table rotates and the soil mix is compacted with a compaction force of 15,000 kg corresponding to a pressure of 4.8 MN/m^2. The manufacturer claims that the dry compressive strength of such treated bricks is between 3.9 and 9.7 MN/m^2. If higher strength bricks are needed, an addition of 1 to 3 per cent cement to the treated soil would result in compressive strengths greater than 10 MN/m^2. It is also claimed that a crew of 4 to 5 workers can produce 3,000 to 4,000 bricks per day from one plant.

The authors have no direct experience of this type of machine but have received favourable comments from Ghana and Malaysia. In Ghana, for example, the Ministry of Works and Housing Test Laboratory tested, in 1977, blocks made on the CLU 3000 brick plant and obtained the following average results for stabilised soil blocks:

Dry compressive strength: 3.46 MN/m^2;
Wet compressive strength: 1.99 MN/m^2.

Figure IV.11

CLU 3000 stabilised soil block making plant

It should be noted that these results are below the figures claimed by the manufacturer.

VII.8 Supertor block making machine

Torsa Maquinas y Equipamentos Ltd. of Sao Paulo, Brazil developed the Supertor block making machine during the 1960s. This company manufactures a range of hydraulically assisted soil-cement block presses. Each machine weighs approximately 1,000 kg and is powered by a 5 hp electric motor. The machines are capable of producing about 20,000 blocks per 8-hour day. One particular model has a mould which can be subdivided to produce 4 blocks in one single pressing, each block measuring 230 mm x 110 mm x 50 mm or 200 mm x 100 mm x 50 mm.

VII.9 Maquina block making machine

This machine was developed during the early 1970s in Bogota, Colombia, and is now widely used in South America. It is a truly local, medium to low-cost machine.[1] It operates on the principle of a pull-down lever, similar to that of the Ellson blockmaster machine. It can exert a compacting pressure of approximately 1.8 MN/m^2.

VII.10 Brepak block making machine

Extensive research was conducted at the Building Research Establishment (BRE) in the United Kingdom during the late 1970s on the production of stabilised soil building blocks. It involved a field study of block-making machines available on the market and extensive laboratory studies on the process of soil stabilisation. One important conclusion derived from the studies is that stabilised soil can be an extremely useful building material for developing countries, provided that an adequate programme of testing is carried out on the raw material. Experimental research carried out in BRE indicates that compacting pressures in the range of 8 to 16 MN/m^2 could give satisfactory and economical results for the production of good quality stabilised soil building blocks.[2]

---

[1] Detailed description of the machine is provided in Roland Stulz, 1981.
[2] See M.G. Lunt, 1980

In the early 1980s, the Oveseas Division of BRE developed a prototype block making machine, referred to as the Brepak machine (see figure IV.12). This machine weighs about 150 kg and produces stabilised soil blocks 290 mm x 140 mm x 100 mm.

Field trials in various parts of the world indicate that about 300 blocks can be produced, on average, during an 8-hour day. A compacting force of about 40 tonnes, equivalent to a compacting pressure of 10 MN/m^2, is exerted by a hand-operated lever hydraulically assisted to produce this pressure. Figure IV.13 illustrates the good-quality blocks that can be produced with this machine.

A joint Anglo-Kenyan research project indicates that large numbers of high-quality blocks may be produced with a Brepak machine. These blocks have the appearance of fired clay bricks and do not need any external rendering to resist the weather.[1]

The Brepak machine is now being used in about 25 countries and is commercially available from Multibloc Ltd., Bristol, United Kingdom (see Appendix IV).

VII.11 Zora hydraulic block press

A simple hydraulic press developed by Zora International Co., Ltd. (United Kingdom) in the early 1980s produces a wide range of stabilised soil blocks. The all-steel press has a mould which can produce building blocks 280 mm long, 125 mm wide and 100 mm thick. The manufacturer claims that this type of press can be operated by unskilled workers and is sturdily built to withstand rigorous outdoor operating conditions with little maintenance. There are three versions, each equipped with hydraulic power supplied from one of the following power sources: a 1 hp electric motor; a 5 hp petrol engine;and manual power. Each model weighs about 800 kg and is fitted with the same basic mould components mounted on an identical two-wheel chassis for easy movement on site.

---

[1] For further details on this research project, see D.J.T. Webb, 1983.

Figure IV.12
The Brepak machine

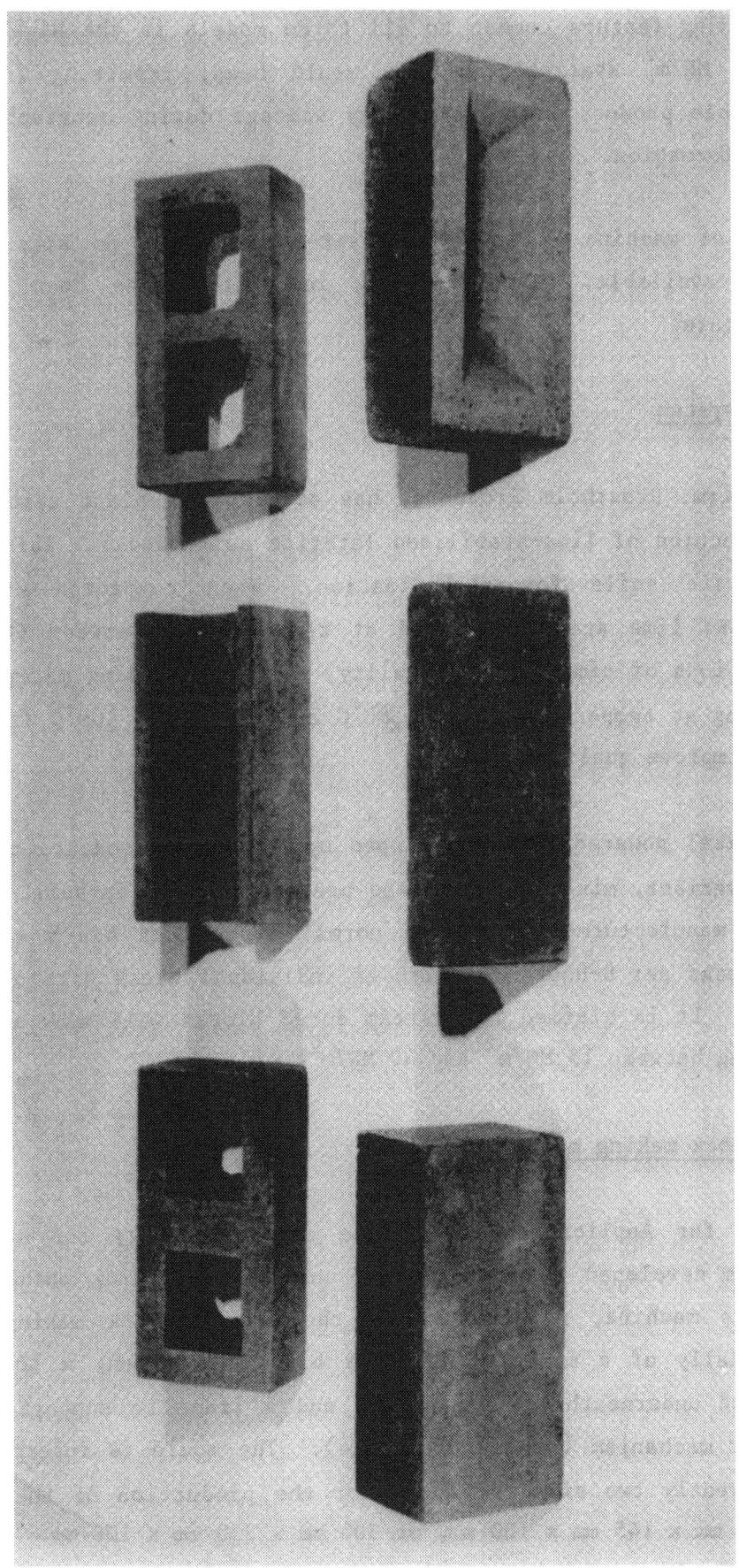

Figure IV.13
Stabilised soil building blocks produced
by the Brepak machine

An outstanding feature common to all three models is the high compacting pressure of 19 MN/m^2 available at the mould head, resulting in a highly compacted, durable product with hardly any wastage during manufacture due to breakage or malformation.

This type of machine is undergoing site trials but no site production rates are yet available. The foregoing information has been taken from existing literature.

VII.12 **Latorex system**

A Danish firm, Drostholm Products, has developed a plant system for the high speed production of lime-stabilised laterite soil blocks. This plant can use only laterite soils for stabilisation. When compacted mixtures of laterite soil and lime are moist-cured at temperatures between 60 and 97°C over various periods of time, a good quality, durable building material can be produced. Curing at temperatures above 80°C and nearer to 100°C for 24 hours should further improve quality[1].

The electrical powered plant developed by Drostholm Products comprises a soil drier, pulveriser, mixing machine and presses with an in-built steam oven for curing the manufactured blocks. A normal size plant has a capacity of about 12,000 blocks per 8-hour day, with an individual block size of 230 mm x 110 mm x 55 mm. It is claimed that steam cured blocks will have compressive strengths varying between 15 MN/m^2 and 40 MN/m^2.

VII.13 **Astram block making machine**

The Centre for Application of Science and Technology for Rural Areas (ASTRA) in India developed a hand-operated soil block making machine in the mid 1970s. This machine, referred to as the Astram block making machine, consists essentially of a mould in which a block is formed, a toggle lever mechanism mounted underneath the mould body and a frame to support the mould and toggle lever mechanism (see figure IV.14). The mould is interchangeable. There exist currently two sizes of mould for the production of the following block sizes: 300 mm x 145 mm x 100 mm, or 300 mm x 230 mm x 100 mm.

---

[1] For more details, see T.C. Hansen and T. Ringsholt, 1978.

Figure IV.14
The Astram block making machine

From a design point of view, the Astram machine looks like a CINVA-Ram press which would be equipped with the toggle mechanism from the Ellson blockmaster without the projecting inclined legs. It can exert a compaction pressure of about 5 MN/m^2. It is stated by the manufacturer that the Astram machine is superior to both the CINVA-Ram and the Ellson blockmaster machine.

VII.14 Tecmor equipment

The manually-operated Tecmor soil cement brick-making machine (model MRC-1)[1] is claimed to produce up to 2,000 bricks per day in two sizes: 230 mm x 110 mm x 50 mm, or 210 mm x 100 mm x 50 mm.

The Tecmor machine looks like the CINVA-Ram machine but has improved vertical guiding to facilitate the compacting load application by the main lever arm. The compaction pressure of 2.5 MN/m^2 is slightly higher than that of the CINVA-Ram machine. This is due entirely to a longer operating lever. The tare weight of this machine is about 85 kg.

Two other types of hydraulic machine are available under the Tecmor trade name: models HRC-1 and HRC-2. They are both powered by 7.5 hp electric motors. Each factory-installed machine has a production rate of 1,500 units per hour. Model HRC-1 is used for the production of two sizes of common bricks: 230 mm x 110 mm x 50 mm, or 210 mm x 100 mm x 50 mm.

Model HRC-2 is used for the production of two sizes of common hollow bricks (230 mm x 110 mm and 210 mm x 100 mm) and one size of solid bricks (510 mm x 230 mm). The above bricks can be produced in various thicknesses varying from 20 mm to 90 mm.

With the factory-installed machines, the company supplies a rotating sieve and a horizontal pan-type mixer which can mix a batch of 200 kg every three minutes. The above equipment can produce soil cement blocks with one part of cement to fifteen parts of soil. The manufacturer claims that, with about 10 to 15 per cent of water, this is the most economical mix for the

---

[1] This machine is manufactured by Equipamentos Meccanicos Ltda. of Brazil (see Appendix III).

production of stabilised soil blocks. However, the manufacturer recommends that tests should be conducted before deciding on a final mix.

VII.15 Meili 60 manual soil block press

The Meili Engineering Company (Switzerland) has developed an improved version of the CINVA-Ram machine, the Meili 60 manual soil block press. This particular machine, which operates on the principle of the off-centre press, is ruggedly built and achieves a compacting force of 20 tonnes, which equates to a compacting pressure of about 5 MN/m^2 when producing 250 mm x 125 mm x 80 mm soil blocks[1]. The manufacturer claims that between 60 and 120 blocks per hour can be produced depending on the size of the labour force employed.

As a result of the successful operation of the Meili 60 soil block making machine in field tests in Guinea, Nigeria and India, the firm developed a power-driven machine, the Meili Mechanpress. It is an automatic soil brick and block making machine based on the original turntable principle used for the Winget rotary table press machine. It is mounted onto a three-wheeled trailer complete with a built-in diesel engine developing 18.5 hp at 2,700 rpm, a horizontal pan type mixer of 150 litres capacity, various moulds and a rotary table press.

The moulds vary from a standard size of 250 mm x 125 mm to a maximum size of 300 mm x 150 mm. The machine can produce one block every 4 seconds. This machine is thus capable of producing about 1,000 high-quality soil blocks per hour. The tare weight of the machine is 1,700 kg.

The authors have no first-hand experience of the above two presses. However, the description in the manufacturer's catalogue tends to indicate that they include a number of improvements over other similar machines.

VII.16 Terrablock Duplex Machine

The Terrablock Duplex trailer-mounted machine, powered by a 43 hp diesel engine, can produce 300 mm x 250 mm x 100 mm adobe soil blocks at a maximum rate of 10 blocks per minute. This process uses wet soil from the ground and a built-in computer controls the fully automatic operation of block manufacturing.

---

[1] For further details on this machine, see SKAT, 1984.

The main hopper holds enough soil for 10 minutes of continuous operation. A heavy duty, built-in sieve filters out debris and oversize particles, while a vibrating device to the hopper head ensures a consistent flow of soil into the block moulds located at the lower end of the hopper.

A horizontally-mounted, double-acting hydraulic ram is employed to compact the soil within a mould. After compaction, the block is automatically ejected from the mould onto a simple conveyor belt.

The manufacturer claims that the operation of the machine is a simple one-man task. As long as the hopper remains loaded with soil, the machine will automatically produce three to five blocks per minute from each of the two moulds. Enough blocks may thus be produced in one hour to construct a 9 m^2 wall, 250 mm thick.

The soil used in this process is _not_ stabilised, and the resulting blocks would therefore be called adobe blocks. It is thus essential to treat a Terrablock wall with a fast drying chemical sealant before applying a finish coat of external rendering to prevent erosion.

The Terrablock adobe block making machine is illustrated in figure IV.15.

VIII. WORLD SURVEY OF SOIL BLOCK MAKING EQUIPMENT

The purpose of this chapter is to draw attention to the various types of forming devices available on the market for the production of stabilised soil building blocks.

The presses described in the previous section and others listed in table IV.1 are obviously not the only ones available on the market. Many other presses produced in both developing and developed countries are currently marketed, but the authors could not obtain information on these presses when this memorandum was being prepared. Additional names of manufacturers and/or suppliers of stabilised soil block making machines are given in Appendix IV, including a very brief description of some of the machines. It must be emphasised, in this context, that the mention of equipment suppliers or manufacturers in this publication does not imply a special endorsement of these by the ILO. The names listed are only provided for illustrative purposes and potential producers of stabilised soil blocks should try to obtain information from as many suppliers as feasible.

Figure IV.15

The Terrablock adobe block making machine

Table IV.1

Survey of soil block making machines

IV.1 A: Block making machines described in Chapter IV

Name	Country of origin	Approx. year introduced	Manual (M) or power (P) operation	Gross weight (kg)	Compacting pressure (MN/m^2)	Max. daily production rate	No of workers	Approx. price (US$,1985)	Maximum block size(mm)
Astram	India	Mid 1970s	M	110	5.0	n.a	3-4	375	300x230 x 100
Brepak	United Kingdom	1979	M	140	10.0	300	3	1,300	290x140 x 100
Ceta-Ram	Guatemala	Mid 1970s	M	80	2.4	250	3	450	290x140 x 90
Cinva-Ram	Colombia	Early 1950s	M	60	2.0	350	3	300	290x140 x 90
Consolid AG	Switzerland	Late 1970s	P	1,600	.8	3,500	6^2	20,000	250x120 x 75
Ellson Block-master	India	Early 1970s	M	210	7.0	750	10^2	---	290x190 x 90
Landcrete/ Terstaram	Belgium	About 1950	M and P	320 2,100	4.0	1,000 2,000	7^2	1,000 18,000	295x140 x 90
Latorex system[1]	Denmark	Mid 1970s	Factory	---	5.0	12,000	---	---	230x110 x55
Maquina	Colombia	Early 1970s	M	170	1.8	180	4	---	200x150 x 60
Meili	Switzerland	Late 1970s	M	120	5.0	500	---	700	250x125 x 40
Supertor	Brazil	Mid 1960s	and P P	1,700 1,000	6.0	7,000 20,000	---	---	230x110 x 80
Tecmor	Brazil	Late 1970s	M and P	85 2,500	2.5	2,000	6	---	230x110 x 50
Tek-Block	Ghana	Early 1950s	M	90	2.0	250	3	240	290x215 x 140
Terrablock	USA	1985	P	5,350	---	4,800	---	80,000	300x250 x 100
Winget	United Kingdom	1948	P	1,100	9.5	1,150	5	---	300x150 x 100
Zora	United Kingdom	1982	M and P	230 850	19.0	---	---	3,000	280x125 x 100

--- = Not available.

IV.1 B: **Block making machines not described in the text**[3]

Name	Country of origin	Approx. year introduced	Manual (M) or power (P) operation	Gross weight (kg)	Compacting pressure (MN/m^2)	Max. daily production rate (8 hrs)	No of workers	Approx. price (US$,1985)	Maximum block size(mm)
La Palafitte	France	1975	M	---	1.4-2.0	240-320	3	---	290x140x90
CENEEMA Earth and Loam Block Press	Cameroon	1979	M	---	---	320-480	3	---	300x140x110
AVM Block Press	F.R.Germany	1984	M	---	---	320-480	3	---	---
SISD Dirt-Cement Brick Press	Thailand	---	M	---	---	320-480	3	---	---
MARO Block Press	Switzerland	---	M	---	---	320-480	3	---	---
CTBI Block Press	France	---	M	85	---	400-720	3	---	290x145x110
UNATA Press	Belgium	---	M	80	---	320-480	3	---	290x140x90
A.B.I. Block Press	Côte d'Ivoire	---	M	---	---	320-480	3	---	---
CTA Block Press	Paraguay	---	M	---	---	600-700	4	---	---

Table IV.1 B (Continued)

Name	Country of origin	Approx. year introduced	Manual (M) or power (P) operation	Gross weight (kg)	Compacting pressure (MN/m^2)	Max. daily production rate (8 hrs)	No of workers	Approx. price (US$,1985)	Maximum block size(mm)
GEO 50	France	---	M	100	---	160-400	2	---	290x140x90
SATURNIA	Switzerland	1983	M	200	---	800-1,200	3	600-1,000	---
RIFFON Block Press	Belgium	---	M	150	---	800-960	3	---	220x105x60
CRATERRE PEROU Block Press	Peru	1982	M	230-280	1.5-2.0	800-960	5	---	280x280x80; 280x128x80
CERAMAN Manual Press	Belgium	---	M	330	2.1	1,600-2,400	4	---	220x107x70
SEMI-TERSTAMATIC	Belgium	1953	M and P	765-925	---	2,500-5,000	---	---	220x105x60; 295x140x90
CERAMATIC Automatic Brick Press	Belgium	1953	P	1,650	6.3	12,000-16,000	2	---	220x107x70
LESCHA SBM	F.R.Germany	1976/84	P	---	8	5,600	4	---	250x130x75

Table IV.1 B (Continued)

Name	Country of origin	Approx. year introduced	Manual (M) or power (P) operation	Gross weight (kg)	Compacting pressure (MN/m^2)	Max. daily production rate (8 hrs)	No of workers	Approx. price (US$,1985)	Maximum block size(mm)
ECOBRICK 1000	Switzerland	1984	P	600	3-10	800	2	---	250x120x75
TERRE 2000 Presse TMR6750-40	France	1984	P	1,800	9	2,400	---	---	300x150x150
GEO 500 Semi-Bloc	France	---	P	---	---	1,350	2	---	295x140x90
ULTRABLOC IMPACT 1 and 2	USA	---	P	1,000-1,200	---	1,700-2,400	---	---	305x140x90
TERRA BLOCK Duplex	USA	---	P	3,700-	5-8	2,800-4,800	4	---	305x140x90
Lorev	Italy	---	M	150	3.0	---	---	---	300x150x60
PPB Saret (Teroc)	France	---	P	---	---	800	---	---	---
Raffin	France	---	M	---	2.5	300	4	---	260x130x80

1 The Latorex blocks are steam-cured whilst all the others are atmospherically cured.

2 Estimates include labour for soil preparation and mixing.

3 The information contained in Table IV.1.B is provided for illustrative purposes only. The productivity and other data shown in this table have not been checked for accuracy by the ILO. The reader is therefore urged to obtain additional information from the manufacturers listed in Appendix III.

--- = not available.

CHAPTER V

CURING AND TESTING

I. INTRODUCTION

Various natural building materials (e.g. wood, straw, foliage, soil) are used in developing countries for the building of what is often considered sub-standard housing of a temporary nature. The same view applies to low-cost urban housing built with a large variety of waste materials, such as scrap metal, cardboard, and so on. This being the case, the proponents of stabilised soil blocks emphasise that good-quality blocks should be considered as durable and as protective (against hot or cold weather, rain, wind, etc.) as building materials such as concrete blocks, fired bricks or building stones jointed with cement-based mortars. It is necessary, however, to choose carefully the materials for the manufacturing of stabilised soil blocks, to apply appropriate soil preparation and forming techniques, and to ascertain that freshly produced stabilised soil blocks are properly cured. The production of good-quality blocks also requires careful testing of the raw materials, especially soil, as well as testing of the output in order to ensure that blocks of the right quality standard will be marketed. The purpose of this chapter is to describe the various curing and testing procedures now available. These procedures should help to improve the quality of the blocks produced and to minimise the probability of marketing sub-standard or defective blocks.

II. THE NEED FOR CURING AND TESTING

A building structure is subjected to various forms of loading which fall into three distinct groups:

- a static dead loading which is always present, made up of the self-weight of building components used within the structure plus the internal fixtures and fittings;

- a live loading caused by the vibration effects of moving loads within the structure (usually taken as an added factor of the induced or dead-loading figure); and

- a dynamic loading caused by the application of external forces such as might occur in a natural hazard (e.g. wind or seismic forces). The dynamic loading is taken into consideration in the design of the building. In this case, the architect must refer to local codes of practice, which usually include a safety factor.

The above three types of expected loading are then combined and a detailed analysis determines the individual strength requirement of each of the different building elements used within a structure.

In the United Kingdom, compressive strengths of 2.8 MN/m^2 are the minimum requirements for concrete blocks and fired clay brick products.[1] Legal building code requirements are usually clearly spelled out according to the nature of the material and its state at the time of testing (e.g. wet or dry). When considering the wet compressive strength of a concrete block, there is a small drop in strength from dry to wet conditions of about 10 per cent. There are numerous factors which may affect compressive strengths including the type of mix and aggregate used, the method of curing after initial casting, and so on.

The percentage drop in strength from dry to wet conditions for fired bricks is higher than for concrete blocks and can, on occasions, be as high as 20 per cent. This factor also depends upon the type of raw material used, mixing, drying and firing conditions, and other environmental variables.

In the case of stabilised soil building blocks, tests show that a drop of about 50 per cent between dry and wet strengths may be expected. The strength of a stabilised soil block depends on several factors, including the type of soil used, the amount of stabilising agent employed, the compaction pressure, the method of curing and the method of testing for strength. Consequently, it is of paramount importance that detailed consideration be given to the method of curing and the procedure of structural strength determination in order to ensure that good-quality blocks are produced and marketed.

---

[1] See British Standards Institution BS6073, 1981 and BS3921, 1974.

III. METHODS OF CURING

To gain maximum strength, stabilised soil building blocks require a period of damp curing. This is a common requirement for all cementitious materials. As already discussed in Chapter II, various cementitious materials can be employed to stabilise clay-based soils. Therefore, only general guidance will be given in this chapter to ensure that good-quality blocks are produced, since proper curing is not the only factor determining the quality of a block.

Once a freshly moulded block is removed from the block forming device, it is imperative that the moisture of the soil mix is retained within the body of the block for a few days. If the block is left exposed to ambient conditions, the surface material will lose its moisture and the clay particles will shrink. This will cause surface cracks on the block faces.

One method of keeping the block moist is simply to insert the block in a plastic bag. Another effective method consists in placing five or six freshly moulded blocks into a plastic refuse bag or dustbin liner (see background in figure V.1). Caution is necessary to prevent the corners of the blocks from breaking, since they will have little strength while being cured. After the bag has been filled with blocks, its open end should be closed in order to retain any free moisture. Alternatively, freshly moulded blocks can be laid out in a single layer, on a non-absorbent surface, and covered with a sheet (e.g. plastic sheets) to prevent the moisture from escaping.

After two or three days, depending on local temperatures, cement stabilised blocks complete their primary cure. They can then be removed from their protective cover and stacked in a pile, as illustrated in the foreground of figure V.1. If lime is used as a stabiliser, the blocks should be left to cure for about 7 days. Water should be sprinkled on the stack and a cover (e.g. plastic sheet, grass or reeds) placed over the top of blocks.

As the stack of blocks is built up, the top layer should always be wetted and covered, and the lower layers should be allowed to air dry and achieve maximum strength. Figure V.2 shows a stack of curing blocks 1.5 m high.

Figure V.1
Curing stabilised soil blocks

Figure V.2
Stack curing

The duration of curing needed varies from soil to soil and, more importantly, with the type of stabilising agent used. With cement stabilisation, it is advised to cure blocks for a minimum of three weeks. The curing period for lime stabilisation should be of at least four weeks. Stabilised soil blocks should be fully cured and dry before use in a construction project. If it is not the case, cracks will most probably appear either on the blocks or across the joints between blocks.

IV. TESTING STABILISED SOIL BUILDING BLOCKS

Stabilised soil building blocks should be considered as structural elements similar to fired bricks and concrete blocks. It is therefore important to submit them to testing procedures similar to those used for the latter materials.

The production of stabilised soil building blocks is often a rural activity. Therefore, it would be wise to consider site testing procedures in addition to laboratory test methods. In both cases, accurate records must be kept. These should include soil mix details, method of manufacture, block dimensions, age of sample and maximum crushing load.

IV.1 Site testing procedures

It can be both time-consuming and expensive to send stabilised soil building blocks to a laboratory for structural analysis. Therefore, it is wise to use first simple, on-site tests that will give an indication of the suitability of a block as a structural element.

It is usual to test structural components at 28 days, though the same tests can be performed earlier (e.g at 7, 14 and 21 days) in order to determine the strength-time relationship.

Considering the reasonably low strengths developed by stabilised soil building blocks, it is recommended to test them after 28 days. The following simple tests can be carried out on site:

Wet-dry cycling test: Once curing has been carried out, five stabilised soil blocks should be selected at random and completely immersed in water for a period of 12 hours or overnight. They are removed from water and left to dry in the sun during the day. This procedure, whereby the blocks are wetted and dried, is repeated seven times. The total duration of the seven wetting and drying cycles is approximately a full week.

Inspection of the samples should indicate if anything is wrong with the original soil used or the stabilising agent employed. For example, blocks may slake or fall to pieces, crack, flake or even burst, indicating that the mix must be modified or, as a last resort, a different soil should be found. It is advisable, therefore, to produce first several sets of blocks with different amounts of stabiliser in order to determine whether the problems can be corrected by using an appropriate fraction of stabiliser. If the problems still persist, other mixes of soil and/or stabilisers must be tried out.

Water absorption test: This test can be conducted in conjunction with the wet-dry cycling test. Prior to the first water immersion, each block is weighed and, after overnight immersion, weighed again. A simple calculation can then be performed to determine the percentage moisture absorption by weight:

$$\% M_c = \frac{W_W - W_D}{W_D} \times 100$$

where : $\% M_c$ = Percentage moisture absorption
W_W = Weight of wetted sample
W_D = Weight of dry sample

Experience shows that, if a block has less than 15 per cent moisture absorption, it is likely to exhibit good, long-term durability.

Wet-dry density test: Immediately after making a stabilised soil block, it should be weighed and its dimensions noted in order to determine its wet or freshly moulded density.

At day 28, prior to the wet-dry cycling test, the block is again weighed and its dimensions noted in order to determine its dry or fully cured density. As previously mentioned, a block should have a minimum specific density when freshly moulded of 1,870 kg/m^3, although the recommended specific density is 2,200 kg/m^3. The minimum dry specific density of a block should be 1,700 kg/m^3, with a recommended density of 2,000 kg/m^3.

If the measured wet and dry density are lower than the minima specified above, the soil mix should be adjusted and/or the processing method revised.

Ring test: If after the required 28 day cure period, two blocks are knocked together and a good "ringing" sound is heard, the blocks should be reasonably dense and weather resistant.

Compressive strength determination using a CINVA-Ram press: It has been stated earlier that a wetted block 290 mm long and 140 mm wide should ideally have a compressive strength of 2.8 MN/m^2. Testing of the block will, in this case, require the application of a load of 11.6 tonnes. This type of loading is excessive for a simple site compression machine. Therefore, it is necessary to reduce the load needed to crush a block. This may be achieved by testing a smaller square block with 100 mm sides, leaving the block height untouched. The crushing load needed would then be reduced to about 2.86 tonnes. The test block may be cut from a normal size stabilised soil block.

A CINVA-Ram machine, with a 25 kg weight hung on the main operating handle, can exert a maximum vertical compaction force of about 3.25 tonnes, when the handle is in the horizontal position. Such a machine may then be used for the compressive strength test using a block with 100 mm sides and 90 mm or 100 mm thickness. The block is placed into a CINVA-Ram mould and a weight of 25 kg is hung from the machine handle. A good-quality strength block should support the load applied without crushing when the handle of the press is in the horizontal position.

If other types of block pressing machines are available, the resulting compacting pressure would have to be determined.

Compression test with a simple lever mechanism: A wetted stabilised soil block sample of 100 mm x 100 mm x 90 mm or 100 mm thick is placed under an operating arm and two men weighing a total of 140 kg sit on the seat provided at the end of the operating arm. A good-quality strength block should withstand the force applied. The apparatus used for this test is shown on figure V.3. It can be manufactured locally.

An alternative, simple machine to carry out dependable compression tests on site has been developed by the United States Department of Housing and Urban Development. This machine can crush 2-inch diameter and 2-inch high cylinders made from unstabilised soil. A slight modification to this machine,

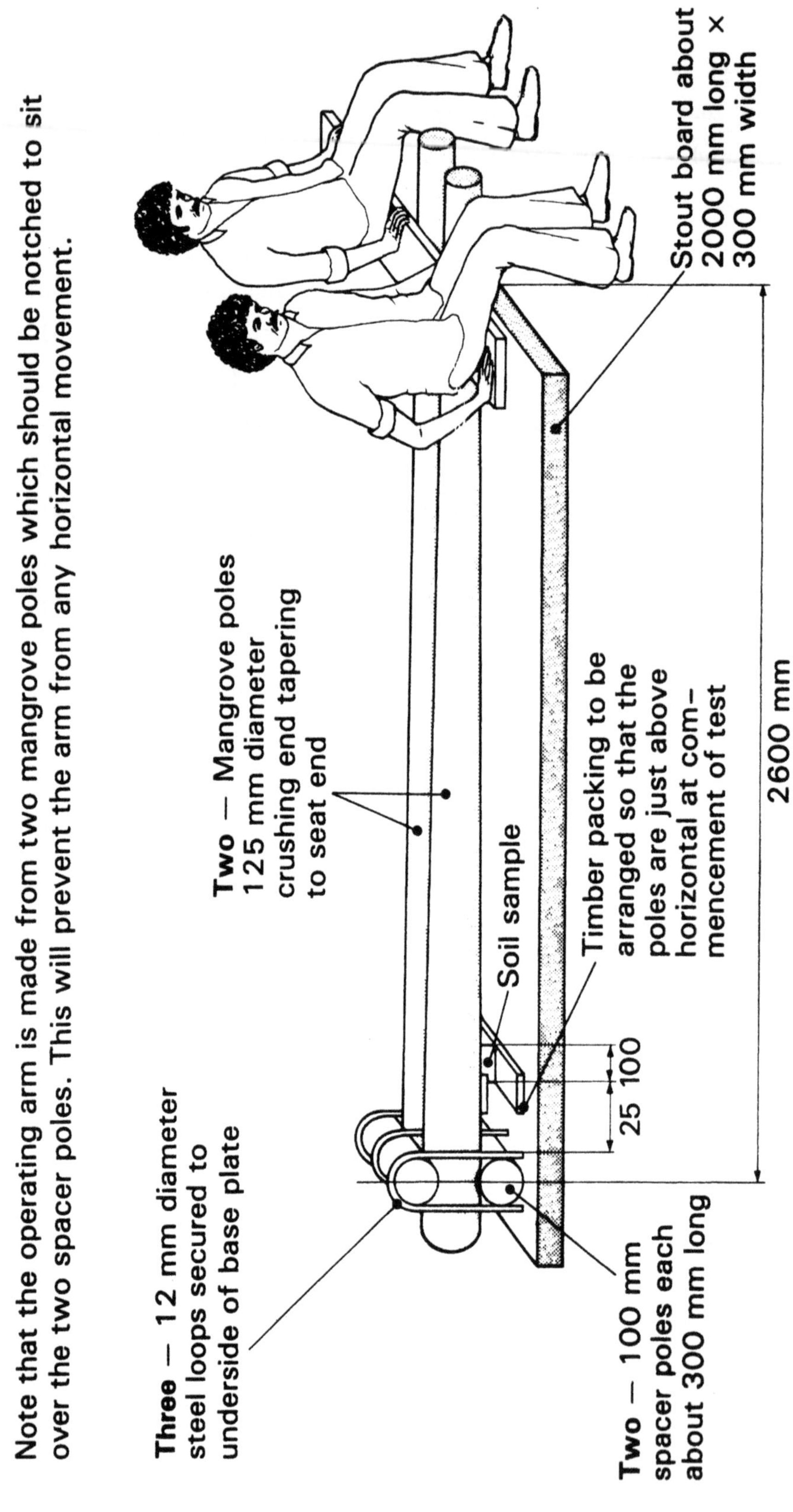

Note that the operating arm is made from two mangrove poles which should be notched to sit over the two spacer poles. This will prevent the arm from any horizontal movement.

Three — 12 mm diameter steel loops secured to underside of base plate

Two — Mangrove poles 125 mm diameter crushing end tapering to seat end

Soil sample

Timber packing to be arranged so that the poles are just above horizontal at commencement of test

Two — 100 mm spacer poles each about 300 mm long

Stout board about 2000 mm long × 300 mm width

2600 mm

25 100

Figure V.3
Simple site lever crushing machine

as shown on figure V.4, would allow testing of 50 mm cubes of stabilised soil. The test is carried out in the following manner:

A 50 mm cube of stabilised soil, cut from a solid block, will be crushed by a force of about 710 kg, if it is to have a working stress of 2.8 MN/m^2. This characteristic determines the testing procedure. After soaking the sample in water for 24 hours, the centre of the sample is placed at 85 mm from the beam pivot of the testing machine (see figure V.4). The timber beam, which has a total length of 1.8 m and a cross section of 100 mm x 50 mm, weighs about 5 kg. Let W be the weight of the bucket (filled with either sand or water) suspended on the beam. If moments about the beam pivot point are taken:

Cube crushing force x A = W x (distance between pivot and bucket suspension point) + (self weight of beam x distance from pivot to centre of gravity of beam)...........Equation (1)

where W = weight of bucket plus content (sand or water);
  A = distance between pivot of the beam and point of application of the force on the cube = 85 mm
  Cube crushing force = 710 kg;
  Distance between pivot and bucket suspension point = 1,676 mm;
  Self-weight of beam: 5 kg
  Distance from pivot to centre of gravity of beam = 838 mm.

Using the above figures in equation 1 gives:

$$710 \times 85 = (W \times 1{,}676) + (5 \times 838)$$

Solving equation 1 gives W = 33.5 kg. This value of W indicates that the block has the required working stress of 2.8 MN/m^2. A lower value will indicate a lower crushing force and therefore a lower working stress.

The advantage of using a machine of this type is that the applied weight W can be slowly increased and, by means of the above equation, an approximate crushing load or stress can be determined for the stabilised soil sample. If, for example, the resulting blocks are to be used in a dry climate, the wet compressive stress can be lowered from 2.8 MN/m^2 to 2.0 MN/m^2; this is equivalent to a crushing force of approximately 507 kg and to a value of W of

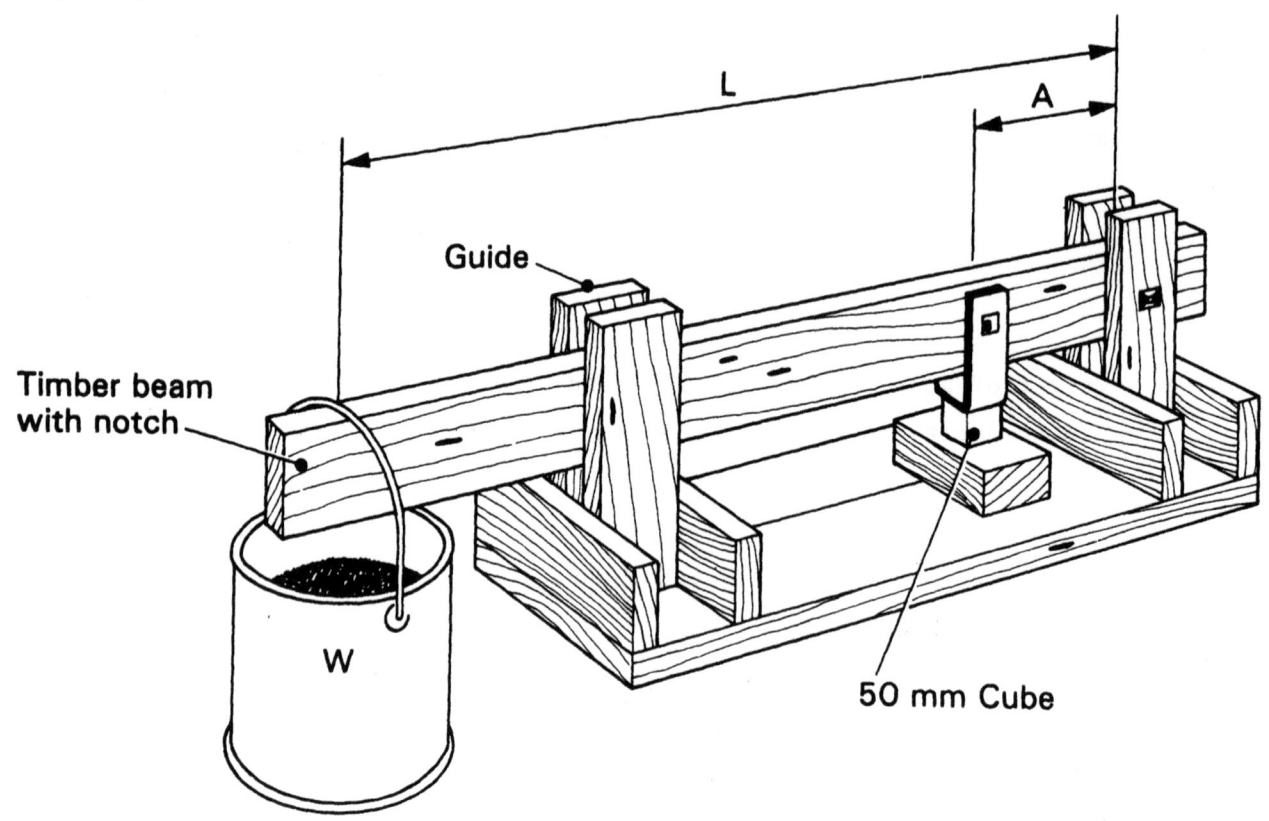

Timber beam : 100 x 50 x 1,800 mm length

A = 85 mm

L = 1,676 mm

W = Weight of hanger and contents

Figure V.4

Site compression test machine for 50 mm cubes

23.2 kg. In practice, the producer must first obtain information on local requirements for wet compressive stress. Then, the value of W is calculated on the basis of equation 1, as long as the testing equipment is identical to that shown in figure V.4. Otherwise, the equation must be revised to take account of different characteristics of the testing equipment (e.g. different lengths of the various components of the equipment, different sample sizes). The value of W is obtained from the relationship:

$$W = \frac{\text{Crushing force in kg} \times 85 - (5 \times 838)}{1,676} \quad \text{......Equation (2)}$$

In the above equation, the crushing force is calculated on the basis of the adopted compressive stress. As an approximation, one may use the following relationship between the crushing force and the adopted compressive stress:

250 kg crushing force applied on a cube with a 50 mm side = 1 MN/m^2

For example, if the local requirement for compressive stress is 3.5 MN/m^2, the crushing force should be equal to 875 kg (250 kg x 3.5). If equation (2) is used, the value of W is then equal to:

$$W = \frac{(875 \times 85) - (5 \times 838)}{1,676} = 41.9 \text{ kg}$$

A value of W lower than 41.9 kg will therefore indicate that the compressive stress of the block is lower than required by local building regulations. This decision must be taken by a qualified engineer who knows the local climatic conditions and possible existing regulations.

These tests will give only approximate results. If more accurate estimates are needed, the blocks must be submitted to laboratory tests, possibly far from the production site, and at a higher cost. The increased accuracy level must therefore be justified before deciding whether to carry out such tests.

IV.2 **Laboratory testing methods**

When using the facilities of a laboratory, it is wise to obtain both the dry and wet compressive strengths of a stabilised soil block. As previously mentioned, there will always be a drop of strength from dry to wet conditions. Thus, it is advisable to ensure that the capacity of a machine is large enough to crush the strongest block.

Ten blocks should be tested for each set of conditions. The time taken before failure of the block should occur within 0.5 to 1.5 minutes for blocks with an expected strength higher than 7 MN/m^2. For compressive strengths lower than 7 MN/m^2, the time taken before failure of the block may be increased to between 1 and 3 minutes.

A crushing load must be continuously applied without shock to the sample at a rate of 5.0 0.5 MN/m^2 per minute on blocks whose expected crushing strength is less than 7 MN/m^2. For blocks with an expected crushing strength above 7 MN/m^2, the loading rate can be increased to 10 1.0 MN/m^2. Other convenient rates of loading up to 35 MN/m^2 per minute could be applied. However, once half the expected maximum load has been applied, the rate should be adjusted to 15 MN/m^2 per minute and maintained until the maximum crushing load is reached.

The above two standards have been developed for known products (i.e. concrete blocks and fire clay bricks). Since stabilised soil blocks fall between the two conditions stated, experience has shown that a load application within the range of 200 kN and 300 kN per minute is acceptable.[1] Ideally, this loading application should be automatically controlled. A suitable compression testing machine would have two ranges: a 3,000 kN range with 1 kN increments and a 750 kN range with 0.1 kN increments. The upper platen of the machine should be attached to a double ball seat mounting. A suitable machine that would comply with the above specifications is illustrated in figure V.5. This machine can be used for a variety of tests. The unit on the left is the main compression test frame, while the unit on the right is used for transverse testing and the application of small loads. Figure V.5 also illustrates a 50 mm diameter cylindrical stabilised soil specimen after crushing on the transverse testing side of the machine.

Figure V.6 shows how a stabilised soil block 290 mm x 140 mm x 100 mm fails when subjected to loading in the main compression frame of the machine. It should be noted that compression test specimens should be "capped" during the test; the block shown on figure V.6 is capped with fibre board.

---

[1] 1 kN = 1,000 N = 0.001 MN.

Figure V.5

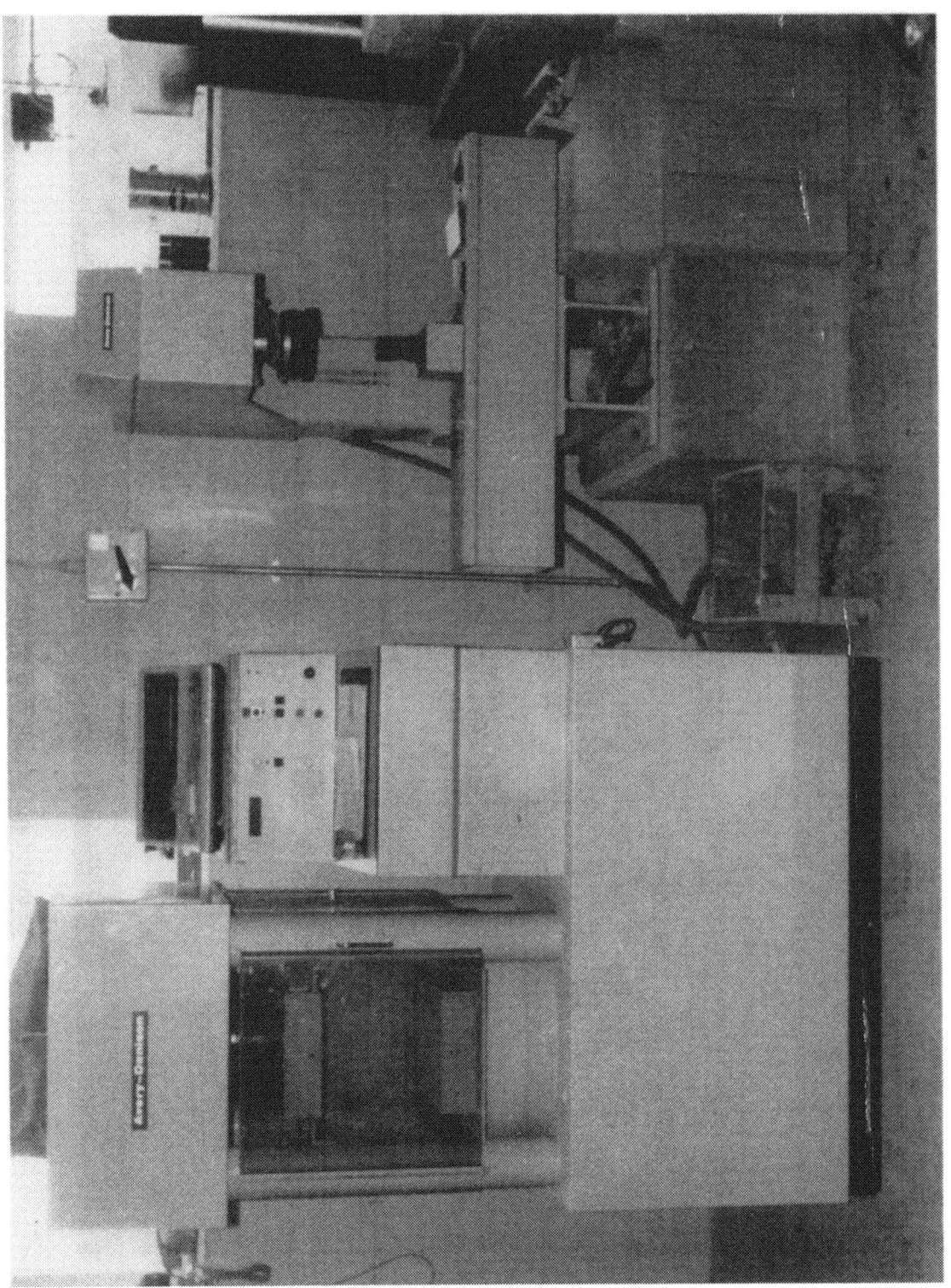

Laboratory compression testing machine

Figure V.6

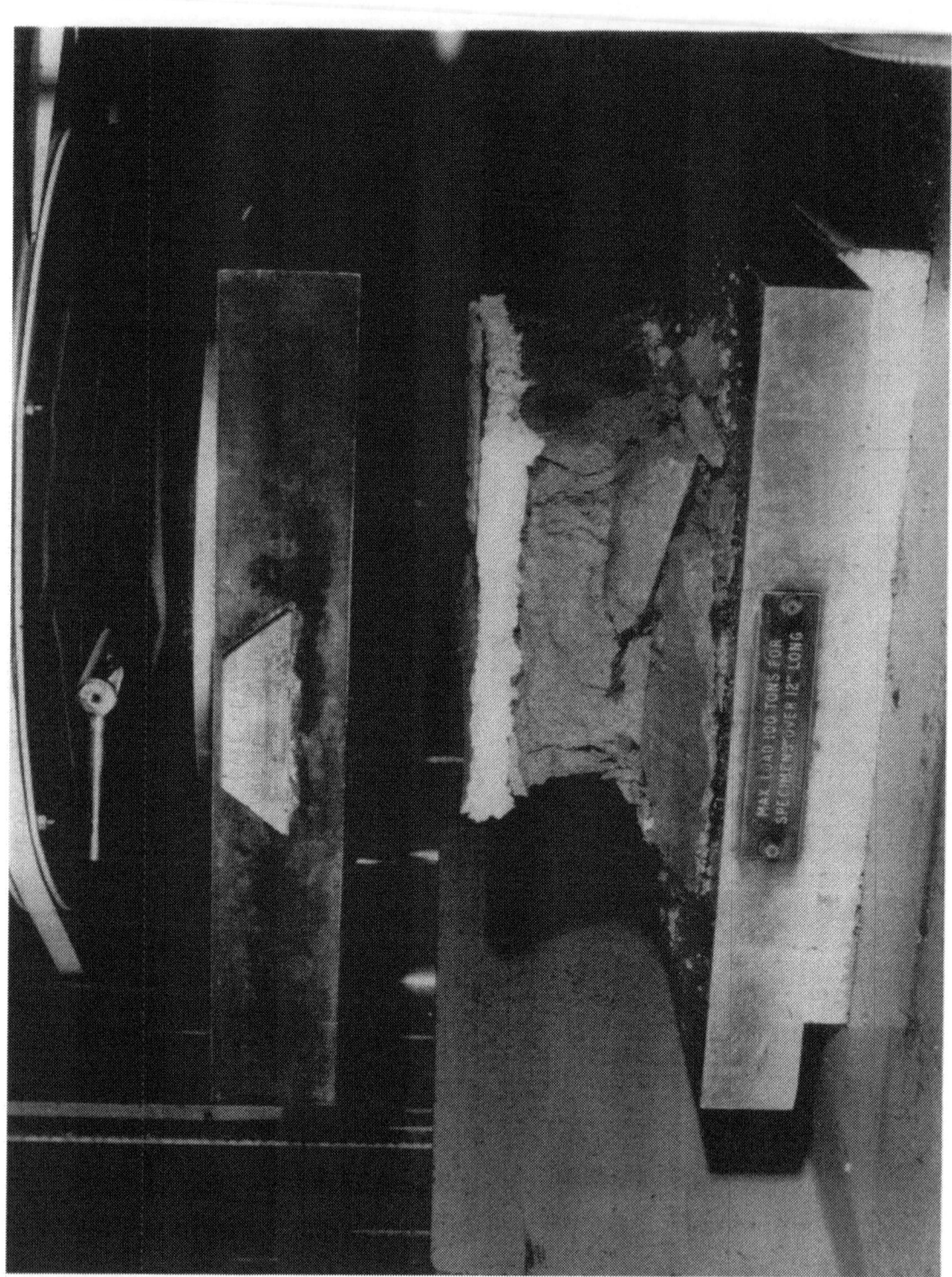

Stabilised soil block after compression test

After the maximum crushing load has been obtained, the crushing strength is determined by dividing the crushing load by the cross sectional area of the block.

IV.3 Durability tests

Two laboratory tests indicate the long-term durability properties of stabilised soil blocks: the water-spray test and the abrasive wear test. These two tests are briefly discussed below.

Water-spray test

The water-spray test is a visual test only. It involves the use of a horizontal spray of water from a 100 mm diameter spray head under a pressure of 1.5 kg/cm^2. A cured stabilised soil block is placed 200 mm from, and parallel to, the face of the spray head. Water is sprayed continuously for 2 hours onto the block, which is then examined visually for erosion and pitting. Test results are indicative only and slight erosion and pitting should not be interpreted unfavourably. Figure V.7 shows the effect of water spray test on a CINVA-Ram manufactured block made from lime stabilised soil. This block, which was manufactured under a compacting pressure of 2 MN/m^2, shows considerable erosion and pitting after a 2-hour spray test. The same soil, also stabilised with lime, but subjected to a compacting pressure of 8 MN/m^2, is much less affected by the 2-hour spray test (see figure V.8). This block is about 10 per cent heavier than that shown in figure V.7. It may therefore be concluded that the strength or durability increases as the dry density increases.

Abrasive wear test

The external face of a building will always be subjected to harsh weather conditions. The spray water test attempts to simulate conditions in rainy and humid areas. In dry arid areas, erosion could occur under the abrasive action of wind-borne sand. Several types of brushing test have been developed to duplicate these conditions. They involve the application of a specified number of brushings with a wire brush, nylon brush heads, or other abrasive materials. However, no firm recommendations have been specified or established for the determination of the resistance of a stabilised soil block to the abrasive action of wind-borne sand.

In the early 1950s, the British Standards Institution introduced a new test for chemical stoneware to determine resistance to abrasion. In this

Figure V.7
Effect of a spray test on a block
compacted at $2MN/m^2$

Figure V.8
Effect of a spray test on a block
compacted at $8MN/m2$

test, eight block samples (each 100 mm x 150 mm x 25 mm) are reciprocated horizontally underneath eight static hoppers containing a fine sand. The sand rubs against the upper face of each sample (which is in contact with the lower face of each hopper) giving rise to an abrasive action. At frequent intervals, each sample is weighed to determine the amount of material abraded away. This weight loss is a good indication of the abrasion resistance of a product.

Figure V.9 shows the results of an abrasive test run on various unstabilised and stabilised soil samples, as well as a sand/cement mortar sample. It gives a clear indication of the relative abrasion qualities of different samples and substantiates the fact that stabilised soil products have good abrasion characteristics.

IV.4 Long-term exposure tests

Many building materials, including stabilised soil products, undergo physical and chemical changes when exposed out-of-doors for a long time. Various short-term site and laboratory tests (such as those described earlier in this chapter) have been devised to simulate the action of weather. However, long-term exposure tests, in which stabilised soil building blocks of known origin and composition are subjected to natural climatic conditions, are a more reliable method for assessing the durability of these materials. Unfortunately, few such long-term exposure tests have been carried out on stabilised soil blocks as they are a relatively new building material.

The Building Research Establishment (United Kingdom) set up a weathering site in 1925 so that various buildings materials could be tested for long periods of time under natural climatic conditions. Figure V.10 shows this exposure site, situated near Watford in the South-East of the United Kingdom.

An experimental building constructed from cement stabilised soil bricks was erected in 1950 on this exposure site (in the background of figure V.10). It was reported in 1974 that the building was still serviceable after 23 years of use.[1] Recently, this experimental building, now used as a general store, was inspected again. It still shows very little deterioration. This indicates that, even in temperate climates, stabilised soil blocks can have a long service life of at least 50 years.

---

[1] See R.G. Smith, 1974.

Figure V.9
Results of an abrasive test on various samples

Figure V.10
The exposure site at Building Research Establishment, United Kingdom

It is desirable to set up a long-term exposure site in countries wishing to expand the use of stabilised soil blocks in order to determine the long-term effects of natural climatic conditions. This would also provide an opportunity to test other forms of soil construction (e.g. adobe, wattle and daub and rammed earth) in order to compare the relative merits of one type of construction against another.

When the CINVA-Ram and Tek-Block machines were being investigated in Ghana (in the late 1970s), a small exposure site was built for long-term exposure tests (see figure V.11). Preliminary results from this exposure site indicate the need for the external protection of some forms of earth construction.

IV.5 Selection of an exposure site

An exposure site should preferably be sited on flat, open ground. If it has to be near trees or buildings, it should not be significantly sheltered from sun, prevailing wind or rain. Sometimes, an exposure site may be used for assessing a variety of building materials. If these include organic materials (e.g. plastics, paints or wood), the site should face the sun (i.e. north-facing in the southern hemisphere and south-facing in the northern hemisphere) in order to irradiate test samples with the maximum of ultra-violet rays. Stabilised soil blocks laid on a flat surface and test walls, such as those shown in figure V.11, should be orientated in such a way as to be subjected to a maximum heating effect of the sun. They should also present maximum exposure to wind and wind-driven rain.

The exposure site should be safeguarded against trespassers or any other form of interference.

After the installation of samples and wall panels, periodic inspections should be carried out to monitor the rate of erosion and general degradation. Results must be recorded for later reference.

For the duration of exposure tests, measurements of meteorological conditions should either be made at the exposure site or be available from an existing meteorological station which is near enough to have similar weather conditions. Comparisons with observations on other exposure sites may then allow some correlation between durability of samples and prevailing local weather conditions.

Figure V.11

Exposure site in Ghana where soil block walls are being weathered

CHAPTER VI

MORTARS AND RENDERINGS

I. NEED FOR MORTARS AND RENDERINGS

Mortars are used primarily to accommodate slight irregularities in size, shape and surface finish of blocks, thus providing accuracy and stability to a wall. In so doing, gaps between blocks are also closed, thus excluding wind and rain from passing through the wall. The mortar has a further advantage in that it improves both the shear and compressive strengths of a wall. Mortars commonly have some adhesive characteristics which improve the shear resistance but do not add significantly to the tensile strength of a wall. In general, mortars need not be stronger than the building blocks.

Rendering applied to the external surface of walls can help prevent ingress of rainwater into a building, and is often used to cover uneven blocklaying. However, if blocks are of good quality and blocklaying techniques are correct, rendering is usually unnecessary. Rendering is preferred in some countries for mainly aesthetic reasons while bare, fair-faced work is preferred in others.

Renderings should be well mixed to minimise shrinkage cracking, then applied to the prepared wall to ensure good surface adhesion. It is advisable to apply a second coat of render, once the first coat has dried and shrunk.

In principle, a good average mix should contain four to five parts sand to one part binder, such as ordinary Portland cement. If insufficient cement is used, the wet mix will be less workable and less strong when set. If too much cement is used, greater shrinkages will occur, the risk of cracking will increase and the cost will be unnecessarily high.

Mortars for laying stabilised soil blocks are often made from the same mix proportions of soil and stabiliser as those used for making the blocks. Many other options exist for the choice of materials used for making mortars and renderings.

II. MORTAR TYPES

There is a large number of mortar types used in the construction industry. The main types are briefly described in this section.

Mud

The most elementary mortar, mud, is made from soil mixed with water. It may be suitable for laying adobe, but is not recommended for stabilised soil blocks. Mud mortar exposed to the weather in fair-faced work will quickly be eroded by wind-blown sand and rain. A good-quality pointing or rendering is essential if mud mortar is used. However, cement used in rendering is preferable for making a more durable mortar.

Soil mixed with stabiliser

The soil and stabiliser used for making blocks may be used in similar proportions for the mortar in which to lay them. As explained earlier, the addition of asphalt or bitumen as a cut-back or emulsion makes a soil more water-resistant. Such mixes can be used as mortars.

Lime and sand mixes

Many countries use lime and sand mixes as rendering materials or mortars. Lime varies in purity and thus gives different types of mortar. If the lime is very pure, consisting of a large proportion of calcium hydroxide, the hardening of the mortar will be due solely to carbonation caused by the slow reaction with carbon dioxide from the air. On the other hand, limes are often impure and contain a proportion of siliceous material from clay contained in the limestone. In this case, lime burning yields a hydraulic lime which will set under water, if necessary. The hardening is, in this case, caused by a pozzolanic reaction between silica and calcium hydroxide which gives calcium silicate. Hydraulic limes also carbonate in air. This type of mortar can be fairly good, but slow hardening makes it less attractive than cement mortars. Replacement of some lime by cement gives a useful increase in early strength.

It is essential that lime be completely slaked before use. Lime is often marketed in a slacked form. The quality of the latter is often satisfactory. Alternatively, quicklime can be used, but it must first be slaked by mixing with water in a pit. A slight excess of water should be added to the quicklime and the mixture covered to prevent drying. Several days in the slaking pit are needed for the complete hydration of quicklime. If any particles of unslaked material remain, they may slake later, after setting. Since slaking is accompanied by expansion, this may spoil the mortar or render.

Lime is often used together with ordinary Portland cement, and mixed with sand.

Pozzolime

Naturally occuring volcanic ashes may contain siliceous material which will react with lime. In the United Republic of Tanzania, three parts of ash and one of lime are mixed together to form a cementitious material. The latter is then mixed with sand for the production of mortar.[1]

Rice husk ash cement

Rice husks burnt at temperatures below 750°C yield approximately 20 per cent of their weight in pure ash. A cement-like material can be produced by mixing two parts by weight of the powdered ash with one part of lime.[2] One part of this rice husk ash cement may then be mixed with three parts of sand (by volume) and water for the production of mortar or rendering. Alternatively, one part of ash may be mixed with two parts of ordinary Portland cement in the production of a Portland/pozzolanic cement.

A similar material can be made from rice husks and lime sludge waste derived from the sugar or paper industries.

Brick-dust/lime mixes

Brickmaking clays fired to only 700°C produce a pozzolanic material when crushed to a fine powder. Two parts of the latter (known as surkhi in India) are mixed with one part of lime for the production of a cement-like material. The latter may then be mixed with sand for the production of mortar.

---

[1] For more details, see R.J. Spence, 1980.
[2] For more details, see R.G. Smith, 1983.

Ordinary Portland cement/sand mixes

Ordinary Portland cement (OPC) mixed with sand in various proportions is widely used for mortars and renderings. Excessively strong mixes may be harmful and unnecessarily expensive. There could be a risk of some spalling of the edges of block faces, and any cracking will probably go through the blocks themselves if shrinkage movement takes place. A weaker mortar might yield a little more under stress, thus reducing the risk of spalling. Furthermore, any cracking would probably go through the mortar joints which are easier to repair than the blocks.

Mixes of ordinary Portland cement and sand are made more workable if lime is substituted for some of the cement. The wet mix is then buttery and easy for the masons to spread. In addition to this beneficial improvement in plasticity, the set of the mortar is retarded, reducing the risk of flash set which may occur in hot climates.

Pulverised fuel ash

Pulverised fuel ash from modern coal-fired, electricity generating plants exhibits pozzolanic properties. Thirty per cent of this material may be mixed with 70 per cent of OPC for the production of pozzolanic cement. Pulverised fuel ash may also be used with lime. The Indian standard specification[1] for this and other pozzolanic materials requires that mortar cubes made up from one part pozzolanic mixture and three parts of sand by weight show an initial set in no less than two hours, and a final set within one to two days, depending on grading. For three selected grades of materials, 28 days compressive strengths should be at least 4.0 MN/m^2, 2.0 MN/m^2 and 0.7 MN/m^2 respectively.

Plasticisers

Instead of adding lime to OPC, very small amounts of purpose-made vinsol resins may be added. The latter form minute bubbles during mixing. However, it is difficult to obtain a mortar with the required properties since the mixing operations cannot be easily controlled. Factors affecting the properties of the mortar include the amount and hardness of water in the mix,

---

[1] See Indian Standards Institution, 1967.

cement quality and quantity, grading of the sand, and the efficiency and duration of mixing which is best done in a high speed mixing machine. Unless all these factors are properly controlled, problems may arise. For example, the mortar may squeeze out after several courses have been laid and the blockwork may get out of vertical alignment if too much air is incorporated in the mix. Furthermore, strength may decrease and the mortar may become too permeable to water after setting. On the other hand, too little air will reduce the mortar workability. Thus, spreading of mortar may not be carried out properly, resulting in low strength blockwork and poor resistance to rain penetration.[1]

Ready-made masonry cements (e.g. cements made with OPC, finely ground mineral filler and plasticisers) are available in some locations. Those which comply with British Standard Specifications[2] can produce strong mortars. Thus, mixes should not contain excessively large proportions of such cements.

Gypsum plaster

Any substance which sets from a fluid into a solid may be regarded as cementitious. Gypsum plasters, made by heating naturally-occuring (or industrially produced) gypsum will set quickly. However, because they are slightly soluble in water, they are not suitable for exterior use in wet climates. Gypsum plaster is mostly used for interior wall finishes.

III. TYPES OF RENDERING

Strong renderings are more likely to shrink and crack than weaker ones. Cracks in rendering result in the moistening of blocks by rain, and therefore in slowing down the drying of walls.

Most of the material types used for the production of mortars may also be used to produce renderings. However, gypsum is not suitable for outdoor rendering in wet climates. Mud renderings may be made more weather-resistant by the incorporation of cow dung. A thin paste, made by adding water to a mix of one part cow dung with four to five parts of soil may also be used to wash over a mud rendering.[3] Soils stabilised with lime or cement would not normally be recommended for use as renderings.

---

[1] For more details, see N. Beningfield, 1980.
[2] British Standards Institution, BS5224, 1976.
[3] For more details, see R. Stulz, 1981.

Water resistance of renderings may be improved by bituminous washes. External rendered finishes of cement/lime/sand mixes are fully described in other publications[1].

Properly made stabilised soil blocks will not require a rendering to protect them from the weather.

IV. MIXING AND USE

Dry ingredients of mixes should be measured out carefully. Although weighing may be preferred, gauge boxes are often used to obtain constant proportions by volume. However, if the water content of sand varies, gauge boxes may not provide accurate mixes.

Sand should generally be clean, free of organic material and salts and well-graded. However, a coarse sharp sand is preferable for a first coat of rendering in two-coat work.

The dry ingredients should be mixed thoroughly prior to final mixing with water. Mixing may be done by hand with spades, or in a mortar mixing machine.

Table VI.1 shows the properties of various mixes of ordinary Portland cement, lime and sand. OPC/lime/sand mixes tend to develop a better bond with blocks, and consequently better resistance to rain penetration. Mixes may be used both for mortars and renderings, the choice of mix being made to suit required properties.

Table VI.1

Mix proportions for the production of mortars and renderings

Mix proportions by volume							Typical compressive strength 28 days (MN/m^2)	Ability to accommodate movement
OPC	Lime	Sand	Masonry cement	Sand	With plasticiser (small amounts) OPC	Sand		
1	0.5	4-4.5	1	2.5-3.5	1	3-4	4.5	Least able
1	1	5-6	1	4-5	1	5-6	2.5	↓
1	2	8-9	1	5.5-6.5	1	7-8	1.0	Most able
Source: British Standards Institution, BS5628, 1978.								

[1] See, for example, Building Research Establishment, 1976.

The mix should be used while still fresh, especially if based on OPC. A good mortar will hang on a mason's trowel, then spread easily on the blocks. It may be necessary to kill the suction of the blocks by dipping or splashing them with water, thus preventing a large proportion of the mixing water being instantly pulled out of the mix as soon as it touches the blocks. Similarly, water may be splashed onto a wall with high suction before rendering is started. If much water is sucked out by the blocks, it will not be possible to spread the mix as either mortar or render, and there may also be insufficient water in the mix to allow the hydration reactions to take place properly. For the same reason, it is preferable to avoid working in the full sun, and to keep the work damp for 24 hours to allow curing to take place. On the other hand, if the mix is too wet, it may have higher porosity, greater shrinkage and lower strength, and the appearance of the finished work may be poor.

Wide mortar joints are sometimes necessary if blocks are badly shaped. With well made blocks, joints should not be wider than 10 mm. This will economise on both materials and labour. If the mortar bed is furrowed, the strength of the wall will be reduced.[1] Vertical joints (or perpends) between blocks should be completely filled with mortar to obtain the best resistance to rain penetration, and to ensure structural integrity between the elements.

Renderings should be applied after the wall surface has been prepared either by hacking the blocks or raking out the mortar joints to a depth of approximately 10 mm, and then brushed free of excess dust, to provide a good key. Renderings may be applied in one or two coats, depending upon the required quality of surface finish. The second coat can be used to fill any cracks in the dried first coat, and to improve the finished quality of the work.

---

[1] For more details, see B. Butterworth, 1953.

CHAPTER VII

COSTING

I. <u>VARIATIONS IN COSTS</u>

The information presented in this chapter is intended to assist entrepreneurs, staff of financial institutions, businessmen and government officials to estimate the production cost of stabilised soil blocks with a view to identifying the least-cost technology and scale of production. A methodological framework for the estimation of production costs is described in the following section of this chapter.

It must be emphasised that the cost of manufacturing stabilised soil blocks will vary a great deal from country to country and even from one area to another within the same country. Unit production costs will vary according to local circumstances, including the following:

- availability of soil; whether it is available on site (e.g. as dug from foundation trenches, etc.) or has to be transported to the site;
- suitablity of the soil for stabilisation, and thus the type, quality and quantity of stabiliser required. It may also be necessary to buy sand if the soil has an excessively high linear shrinkage;
- current prices for commodities, especially stabilising agents;
- whether the blocks are to be made in rural or urban areas; size and type of equipment used; and quality required; and
- current wage rates, and productivity of the labour force.

It is important to note that block making can be carried out on a "self-help" basis, where labour costs will be reckoned to be zero. Furthermore, soil is often available at no cost.

The methodological costing framework described in this chapter is illustrated by a case study in which both labour and soil had to be paid for. On the other hand, low costs were involved in soil preparation and stabilisation because the soil did not need any equipment for crushing, and only a low fraction of stabiliser was needed.[1] The calculations in this chapter are to be used as examples only and should be adapted to the particular circumstances prevailing in a given location, using the appropriate wages, input costs, etc.

II. METHODOLOGICAL FRAMEWORK

The methodological framework consists of 12 steps which may be sub-divided into two main parts:

- Determination of quantities of various inputs (steps 1 to 6);

- Estimation of the cost of each input and calculation of unit production costs (steps 7 to 12).

These steps are briefly described in the remaining part of this section.

Step 1

Determination of the number of blocks to be produced in a given time. The number will be a function of market demand, availability of funds, adopted manufacturing technique, etc. Table IV.1 at the end of Chapter IV, indicates the production rates which may be achieved with different types of presses.

Step 2

Estimation of the quantities of material inputs for the selected scale of production. The principle materials are suitable soil, sand (if soil has a high linear shrinkage), stabiliser and water. Some oil, for example used engine oil, will be required as a mould release agent. Guidance on proportions of components is given in Chapter II.

Step 3

List of the equipment required. This will include items for digging and moving soil, preparing soil with crusher or sieving screen, mixing, a device

---

[1] Details on this case study are provided in J.K. Kateregga, 1985.

for moulding blocks, a covered block curing area and an office. Provision should also be made for soil investigation and testing equipment. Chapters II to V provide the information needed for determining the type of equipment and infrastructure required.

The cost of industrial pieces of equipment may be obtained from equipment suppliers and manufacturers (see Appendix IV) or from local workshops in case the equipment can be manufactured locally.

Step 4: List of labour requirements. The productivity of the labour force may not only vary from one country to another, but also from one site to another within the same country. It is necessary to specify the length of the working day, the number of days worked per week and the number of working weeks per year, taking into account an allocation of time for leave of absence, all within any conditions agreed between unions, employers, etc. The level of skill requirements must also be specified; table IV.1 gives some indications on the number of workers required for selected presses and scales of production.

Step 5

Other local services and facilities may be required; these may include:

- land for quarrying soil for blockmaking;
- land for production area;
- land for curing area and storage of raw materials; and
- provision of access to working area for delivery of materials and dispatch of products.

Little land will be required for small-scale production.

Step 6

Calculation of working capital requirements. In addition to funds for the purchase of equipment and land as itemised in the preceding steps, it will be necessary to have sufficient financial resources for the purchase of raw materials and payment of wages for a period of one month, since there can be no income from the sale of blocks until they have been made and cured. If difficulties are anticipated in obtaining any particular commodity, it might be necessary to maintain sufficient stocks for a period longer than one month.

It may also be desirable to utilise some of the first-produced blocks in the construction of the covered area, office, etc., in order to reduce the cost of items under step 3. It will then be necessary to increase slightly the working capital to allow for the number of blocks which are used for this purpose rather than sold.

Step 7

Annual cost of materials identified in step 2 must be calculated. Clay, sand and water are often extremely cheap commodities. Often, the only signficiant part of their cost is that incurred for extraction and transport. The mould release agent will not be required in large quantities; its cost will therefore be low. Used engine oil may be purchased at a very low price or obtained free; in the latter case, the cost of used oil will be limited to that of transporting it to the project site.

Step 8

Calculation of depreciation costs of equipment and buildings. Whatever the type of equipment used, it will have a limited life. An estimate must be made of the annual depreciation costs for separate equipment items. The depreciation cost of buildings must also be estimated. These costs will depend on the initial purchase price, the life of equipment and buildings and the prevailing interest rate. Depreciation costs may be calculated with the help of table VII.1. This table gives the annuity factor (F) for interest rates up to 40 per cent and expected life periods up to 25 years. Thus, if Z is the purchase price of the equipment or the cost of the building, the annual depreciation cost is equal to Z/F.[1] It can be seen from the table that the longer the useful life of the equipment or building, the lower the annual depreciation cost, and the higher the prevailing interest rate, the higher this cost.

---

[1] The annual depreciation cost calculated in this manner assumes a salvage value of equipment and buildings equal to zero. This simplification does not affect results significantly. Those who wish to take into consideration the salvage value may use other formulations of depreciation costs available in the literature.

Step 9

Realistic figures must be obtained for the cost of labour in the area where blocks are to be produced. Local wage levels for different skills must be used and fringe benefits included in the estimation of labour costs.

Step 10

Land has an infinite life, and the area from which soil is obtained may be restored to its original use in some instances. Thus, the annual cost of land may be assumed to be equal to the annual rent of equivalent land. If the land is already owned by the entrepreneur, a hypothetical annual rental rate should be used when estimating the annual land cost, since this is the income he might have obtained by renting it out instead of using the land himself.

Step 11

Working capital raised on loan for the block making project will require an allowance in the annual cost for interest payments on borrowed capital.

Step 12

The unit production cost may be calculated by summing up the separate cost items from steps 7 to 11 in order to obtain the total annual cost; the latter is then divided by the number of blocks produced annually to obtain the unit production cost. Thus:

Total annual production cost = materials costs + depreciation costs + labour cost + land rental + interest on loan; and

$$\text{Unit production cost} = \frac{\text{Total annual cost}}{\text{Annual output}}$$

III. APPLICATION OF THE METHODOLOGICAL FRAMEWORK

The use of the above methodological framework for the estimation of unit production cost is illustrated, in this section, by a real production situation in Kenya.[1] Individual cost items are expressed in Kenya shillings (1983).

---

[1] Further details on this case may be found in J.K. Kateregga, 1985.

Step 1: Annual production of blocks is 66,000 (i.e. 240 per day).

Step 2: Annual materials requirements

Soil

6.5 kg of soil of suitable clay/silt content is needed per block; thus, 430 tonnes of soil will be used annually. Alternatively, if soil had a high shrinkage, it could be partly replaced by sand. Bearing in mind the different densities of soil and sand, 214 tonnes of soil and 196 tonnes of sand would be required, if they were in equal volumes.

Stabiliser

Ordinary Portland cement is used as stabiliser in the proportion of 4 per cent by weight. Thus, 17 tonnes of OPC are needed annually. Since the soil used had 15 per cent clay/silt content, it was very suitable for block making. Otherwise, a greater percentage of cement would have been required.

Water

One litre of uncontaminated water is needed per block. Thus, the annual consumption of water is 66,000 litres.

Oil

One litre of waste engine oil is needed per 250 blocks. Thus, the annual oil consumption is 260 litres.

Step 3 : Required equipment

For soil selection:
- 1 auger;
- 1 set of sieves;
- 4 linear shrinkage moulds;
- bottles.

For winning the soil:
- 2 wheelbarrows;
- 2 picks;
- 4 shovels.

For soil preparation:
- 1 sieving screen (if soil is adequately crushed);
- 1 pendulum crusher, instead of the sieving screen, if the soil needs crushing.

For block production:

- 1 hand-operated blockmaking machine with spares.

Work area:
- for drying soil, sieving, storage, cure: 50 m^2.

Step 4: Labour requirements for blockmaking

Supervision:	1 foreman/technician
Winning soil:	1 unskilled worker
Preparing and mixing:	2 unskilled workers
Block forming:	2 unskilled workers
Curing, stacking:	1 unskilled worker

The above team works 8.5 hours per day, 5.5 days per week and 50 weeks per year (i.e. 275 working days per year). The supervisor may, however, be involved in some other projects; he is assumed to work 250 days per year on this project.

Step 5: Land and access requirements

The land area for quarrying the soil is estimated on the following basis:

- project life : 15 years;
- digging depth : 1m;

Thus, 3,000 m^2 of land are required (0.3 ha).

(If the block making production site moves frequently, the land requirement for quarrying will be negligible).

Land for access and production, including curing and storage areas: 200 m^2.

Step 6: Working capital requirements

Raw materials for one month: one twelfth annual estimate;
Salaries for one month: one twelfth annual estimate.

Step 7: Annual cost of materials

Item	Quantity	Cost
		(Kenyan shillings, Ksh)
Clay	430 tonnes at 43 Ksh/tonne[1]	18,490
Stabiliser, OPC	17 tonnes at 60 Ksh/50 kg bag	20,400
Water	66 m^3 at 0.4 Ksh/m^3	26
Oil	free	
Total annual cost of materials		38,916

Step 8: Depreciation costs

- <u>Initial cost of equipment</u> (Ksh)
  (assumed life: 3 years)
  one Brepak machine   15,000
  one sieving screen[2]   150
  ancillary equipment   5,000

  Total initial equipment cost   20,150

- <u>Building costs</u> (assumed life: 7 years)

  temporary covered area (post and roof)
  50 m^2 at 30 Ksh/m^2 1,500

- <u>Annual depreciation costs</u>

  - <u>For equipment</u> : F1 (3 years, interest rate 14 per cent) = 2.322

---

[1] If sand is used in place of some of the soil, it will cost twice as much per tonne.

[2] If a pendulum crusher had been required, it would have cost 14,000 Ksh.

Annual depreciation cost $= \dfrac{20,150 \text{ksh}}{2.322} = 8,678$ Ksh/year

- For buildings: F2 (7 years, interest rate 14 per cent) = 4.288

    Annual depreciation cost $= \dfrac{1,500}{4.288} = 350$ Ksh/year

Total annual depreciation cost is then equal to:

$$8,678 \text{ Ksh} + 350 \text{ Ksh} = \underline{9,028 \text{ Ksh}}$$

Step 9: Annual labour costs

- For skilled labour : 55 Ksh/day for 250 days per year = 13,750 Ksh
- For unskilled labour:
    22.50 Ksh/day, 275 days per year, 6 workers = 37,125 Ksh

Total annual labour costs: <u>50,875 Ksh</u>

Step 10: Land rental cost

Small-scale units producing stabilised soil blocks are likely to be situated in areas commending low land value or rental, such as agricultural land. Rental value of the latter may thus be used for preliminary estimation of production costs. Taking into consideration a land requirement of 0.32 ha (see step 5), and an annual rental rate of 1,000 Ksh/ha, the annual rental rate may be estimated at:

0.32 ha x 1,000 Ksh/ha = <u>320 Ksh</u>

Step 11: Interest on working capital

From step 7, the monthly cost of materials is:

38,916 Ksh ÷ 12 = <u>3,243 Ksh</u>

From step 9, the monthly cost of labour is:

50,875 Ksh ÷ 12 = <u>4,240 Ksh</u>

Total working capital requirement:

3,243 Ksh + 4,240 Ksh = 7,483 Ksh

Using an interest rate of 14 per cent, annual interest payments on working capital amount to:

7,483 Ksh x 0.14 = 1,048 Ksh

Step 12: Unit production cost

The total annual production cost is equal to the sum of the following cost elements:

	(Ksh)
Materials	38,916
Depreciation	9,028
Labour	50,875
Land rental	320
Interest on working capital	1,048
TOTAL	100,187

For an annual production of 66,000 blocks, the unit production cost is equal to:

100,187 Ksh ÷ 66,000 = 1.52 Ksh

It may be noted that the above unit production cost will vary from country to country and from site to site within the same country. Although the estimation of the above unit cost takes into consideration production conditions in Kenya, special circumstances in some parts of the country could result in the production of higher or lower cost blocks.

At the time this technical memorandum was being sent for reproduction, new information was received regarding the Brepak block making machine. The latter has been modified and it can now produce up to 360 blocks per day. At this higher productivity level, the unit production cost should be reduced to 1.21 Ksh.

Table VII.1
Discount factor (F)

Year	Interest rate (percentage)														
	5	6	8	10	12	14	15	16	18	20	22	24	25	26	28	30	35	40
1	0.952	0.943	0.926	0.909	0.893	0.877	0.870	0.862	0.847	0.833	0.820	0.806	0.800	0.794	0.781	0.769	0.741	0.714
2	1.859	1.833	1.783	1.736	1.690	1.647	1.626	1.605	1.566	1.528	1.492	1.457	1.440	1.424	1.392	1.361	1.289	1.224
3	2.723	2.673	2.577	2.487	2.402	2.322	2.283	2.246	2.174	2.106	2.042	1.981	1.952	1.923	1.868	1.816	1.696	1.589
4	3.546	3.465	3.312	3.170	3.037	2.914	2.855	2.798	2.690	2.589	2.494	2.404	2.362	2.320	2.241	2.166	1.997	1.849
5	4.330	4.212	3.993	3.791	3.605	3.433	3.352	3.274	3.127	2.991	2.864	2.745	2.689	2.635	2.532	2.436	2.220	2.035
6	5.076	4.917	4.623	4.355	4.111	3.889	3.784	3.685	3.498	3.326	3.167	3.020	2.951	2.885	2.759	2.643	2.385	2.168
7	5.786	5.582	5.206	4.868	4.564	4.288	4.160	4.039	3.812	3.605	3.416	3.242	3.161	3.083	2.937	2.802	2.508	2.263
8	6.463	6.210	5.747	5.335	4.968	4.639	4.487	4.344	4.078	3.837	3.619	3.421	3.329	3.241	3.076	2.925	2.598	2.331
9	7.108	6.802	6.247	5.759	5.328	4.946	4.772	4.607	4.303	4.031	3.786	3.566	3.463	3.366	3.184	3.019	2.665	2.379
10	7.722	7.360	6.710	6.145	5.650	5.216	5.019	4.833	4.494	4.192	3.923	3.682	3.571	3.465	3.269	3.092	2.715	2.414
11	8.306	7.887	7.139	6.495	5.938	5.453	5.234	5.029	4.656	4.327	4.035	3.776	3.656	3.544	3.335	3.147	2.752	2.438
12	8.863	8.384	7.536	6.814	6.194	5.660	5.421	5.197	4.793	4.439	4.127	3.851	3.725	3.606	3.387	3.190	2.779	2.456
13	9.394	8.853	7.904	7.103	6.424	5.842	5.583	5.342	4.910	4.533	4.203	3.912	3.780	3.656	3.427	3.223	2.799	2.468
14	9.899	9.295	8.244	7.367	6.628	6.002	5.724	5.460	5.000	4.611	4.265	3.962	3.824	3.695	3.459	3.249	2.814	2.477
15	10.300	9.712	8.559	7.606	6.811	6.142	5.847	5.575	5.092	4.675	4.315	4.001	3.859	3.726	3.483	3.268	2.825	2.484
16	10.838	10.106	8.851	7.824	6.974	6.265	5.954	5.669	5.162	4.730	4.357	4.033	3.887	3.751	3.503	3.283	2.834	2.489
17	11.274	10.477	9.122	8.022	7.120	6.373	6.047	5.749	5.222	4.775	4.391	4.059	3.910	3.771	3.518	3.295	2.840	2.492
18	11.690	10.828	9.372	8.201	7.250	6.467	6.128	5.818	5.273	4.812	4.419	4.080	3.928	3.786	3.529	3.304	2.844	2.494
19	12.085	11.158	9.604	8.365	7.366	6.550	6.198	5.877	5.316	4.844	4.442	4.097	3.942	3.799	3.539	3.311	2.848	2.496
20	12.462	11.470	9.818	8.514	7.469	6.623	6.259	5.929	5.353	4.870	4.460	4.110	3.954	3.808	3.546	3.316	2.850	2.497
21	12.821	11.764	10.017	8.649	7.562	6.687	6.312	5.973	5.384	4.891	4.476	4.121	3.963	3.816	3.551	3.320	2.852	2.498
22	13.163	12.042	10.201	8.772	7.645	6.743	6.359	6.011	5.410	4.909	4.488	4.130	3.970	3.822	3.556	3.323	2.853	2.498
23	13.489	12.303	10.371	8.883	7.718	6.792	6.399	6.044	5.432	4.925	4.499	4.137	3.976	3.827	3.559	3.325	2.854	2.499
24	13.799	12.550	10.529	8.985	7.784	6.835	6.434	6.073	5.451	4.937	4.507	4.143	3.981	3.831	3.562	3.327	2.855	2.499
25	14.094	12.783	10.675	9.077	7.843	6.873	6.464	6.097	5.467	4.948	4.514	4.147	3.985	3.834	3.564	3.329	2.856	2.499

CHAPTER VIII

SOCIO-ECONOMIC CONSIDERATIONS

I. INTRODUCTION

The previous chapters, which are mostly of a technical nature, are of particular interest to small-scale entrepreneurs, extension agents and the technical staff of government agencies concerned with low-cost housing programmes, such as self-help housing schemes. These technical chapters should promote the profitable production of good quality stabilised soil blocks. However, various constraints of a socio-economic nature may prevent or slow down the wide adoption of this building material, especially in low-cost housing programmes. The purpose of this chapter is therefore to indicate the various socio-economic effects which may result from an expansion of the production of stabilised soil blocks with a view to inducing the formulation of policies and measures in favour of such production.

This chapter is mostly intended for government planners, housing authorities and officials from industrial development agencies who are in a position to promote the necessary legislation and programmes for the development of the production of stabilised soil blocks along with that of other building materials.

II. ACCEPTANCE AND APPLICATION

Soil has been and continues to be the most widely used housing construction material. It is cheap, readily available and may be simply formed into blocks or used in _pisé_ construction. It provides adequate protection against hot and cold weather conditions in view of its high thermal capacity and insulating characteristics. In spite of its long proven use, it is sometimes regarded with doubt and distrust, and is often not recognised by authorities as an acceptable, permanent building material. Its chief technical disadvantage - lack of resistance to weakening and erosion by water - may be mitigated by the use of a stabiliser, as described in this memorandum.

In a number of developing countries, housing authorities have formulated building standards which often rule out the use of soil as an officially acceptable building material. These standards are not applied in all cases: they mainly concern medium- to high-income housing and public buildings which require the delivery of a building permit. Thus, soil is mostly used for dwellings which are built without formal authorisation, such as rural housing or uncontrolled low-income housing in urban areas. This restrictive building standard often applies to stabilised soil blocks although they may be more suitable than officially accepted building materials when used according to sound technical practice.

Although there are now some signs of change, whereby stabilised soil may be allowed, it will be necessary in many countries first to convince the authorities of the suitability of this material, especially when compared to unstabilised soil. In order to do so, stabilised soil construction may have to be developed first in the rural areas, where controls are less stringent, or often non existent. In practice, it may be wise to construct some community buildings first, so that the local people can see for themselves the quality and durability of the material, and experience at first hand the conditions which this method of construction affords. Housing may then follow. With proven success in rural areas, not only will the rural people acquire better housing, but controls for urban areas may then be modified by the authorities to allow stabilised soil construction. This would be to the particular benefit of those living in the outlying areas of the big towns and cities where housing conditions need much improvement. Kenya offers an example among others of a country which has modified its building code to include stabilised soil as a building material.

Following research and development work and the erection of a number of buildings, including a medical clinic,[1] the use of good-quality stabilised soil blocks for walling and flooring is now included in the Government's 1985 Low Income Housing Report.[2] This material is to be included in the Kenyan

---

[1] For more details on the Kenyan experience, see J.K. Kateregga, 1982, 1983 and 1985.

[2] See report prepared by the Kenya Ministry of Works, Housing and Physical Planning, 1985.

Building Regulations after the Kenyan Bureau of Standards has developed its own standards and codes of practice for the production and use of stabilised soil blocks.

It may be noted that a number of industrialised countries are reviving the use of stabilised soil blocks and other forms of earth construction. For example, an international centre for the study and promotion of soil-based construction has recently been established in France.[1] Paradoxically, while the use of soil as a building material concerns mostly low-income housing in developing countries, it is mostly associated with middle to high-income housing in industrialised countries such as France or the United States. This shows that the adoption of stabilised soil blocks for high-income housing in developing countries could be achieved through efficient promotion. For example, housing authorities could finance houses made from stabilised soil blocks, for rent to government officials in order to demonstrate the quality, durability and versatility of this material. Such a project would also show that soil-based housing need not be limited to simple one-storey buildings.

III. EMPLOYMENT GENERATION

The generation of productive employment is one of the most important objectives of national development plans in developing countries. Hence, technologies which require more labour per unit of output than other technologies should be favoured, provided that labour is utilised in an efficient and economic manner.

It can be shown that, in general, the small-scale production of stabilised soil blocks (using intermediate technologies) is much more labour-intensive than that of other, similar building materials such as fired bricks or concrete blocks. Table VIII.1 compares labour requirements for the production of equivalent numbers of stabilised soil blocks and fired bricks. Since the standard sizes of blocks are different from that of fired bricks, and since the comparison should apply to the same volume of walling, it is assumed that one stabilised block is equivalent, in terms of volume, to 2.36 bricks. Four brick-making technologies are compared to one single block making technique using the Brepak press. It can be seen from table VIII.1 that the production of stabilised soil blocks is 2 to 18 times more labour-intensive than that of fired bricks, depending on the techniques which are being compared.

---

[1] International Centre for the Research and the Application of Earth Construction (CRATERRE) at Villefontaine (France). See also Appendix II.

It can also be shown that the production of stabilised soil blocks is more labour-intensive than that of other competing materials, such as concrete blocks.

The production of stabilised soil blocks presents other advantages from the point of view of indirect employment generated. Most countries should be able to produce the tools and equipment needed for small-scale production of blocks, using some of the soil preparation equipment and block presses described in previous chapters. In case some of this equipment is patented, licensing for local production could be arranged. Thus, the production of stabilised soil blocks could generate a great deal of both direct and indirect employment. This is less so for other building materials.

Table VIII.1
Comparative labour requirements

Products	Production method	Labour required to make volume equivalent of 240 blocks per day
Stabilised soil blocks	Small-scale, Brepak press	6
Fired clay bricks	Small-scale, traditional manual process	2.5
	Small-scale "intermediate technology"	3
	Soft mud machine and manual handling	1
	Moderately mechanised technique	1/3

Source: R.G. Smith, 1984 and K.J. Kateregga, 1982.

IV. INVESTMENT COSTS AND FOREIGN EXCHANGE SAVINGS

The local production of building materials which requires the import of expensive equipment and/or intermediate inputs (e.g. cement) can severely tax the limited foreign reserves of developing countries. Thus, building materials which do not require such imports should be favoured. This is the case for stabilised soil blocks which, in terms of foreign exchange savings, compare very favourably with sun-dried clay bricks or building stones. As stated earlier, the tools and equipment required for the production of stabilised soil blocks can be manufactured locally; it may only be required to pay a small licensing fee to the foreign patent holder and/or to import some small parts of the equipment which cannot be manufactured locally. The stabiliser may also be manufactured locally, especially if lime is used as a stabilising agent. On the other hand, if cement is used as a stabiliser, it may need to be imported. Finally, unlike other building materials, the production of stabilised soil blocks does not require energy for drying or firing. Thus, it is not necessary to use imported fuel or to aggravate deforestation caused by the use of local wood for firing clay bricks.

The production of stabilised soil blocks does not require large capital investments which, in developing countries, are usually made at very high interest rates. Thus, the establishment of a small-scale block making plant may be afforded by entrepreneurs who cannot obtain or afford relatively large loans from banks or other sources. The amount of land required is usually small compared with that needed for brickmaking. Furthermore, no land is required if blocks are made at the construction site. The cost of block making equipment can also be very low. In some cases, it need not exceed 1,000 US Dollars for a production capacity of 350 blocks per day: acquisition of a press costing approximately US$400 and that of an earth-crushing/sieving device costing less than US$600.

To summarise, both in terms of capital investment and foreign exchange use, small-scale production of stabilised soil blocks compares very favourably with that of other building materials, especially fired clay bricks and concrete blocks.

V. PRODUCTION COST OF STABILISED SOIL BLOCKS AND BUILDING COSTS

A major component of the cost of a house in developing countries is that of building materials. This is particularly true for low-cost housing and self-help housing schemes. In the latter case, labour is provided by the house owner, who needs to buy only the building materials. Thus, the cost of these materials is, from a financial viewpoint, the only major cost faced by the future house owner. It is therefore important to promote the production of low-cost building materials in order to facilitate home ownership by low-income groups and to reduce public investments by housing authorities.

Compared to other building materials, stabilised soil blocks are also a fairly attractive material from the viewpoint of unit production cost and therefore retail price. This can be seen from table VIII.2, which compares the unit cost of stabilised soil blocks with that of concrete blocks in 1981 Kenyan shillings.

Table VIII.2

Unit cost of stabilised soil blocks and concrete blocks

Organisation of production	Average unit cost of block (1981 Kenyan shillings)
a. Self-help, soil on site	0.45
b. Self-help, soil brought in	0.74
c. Paid labour, soil on site	1.23
d. Paid labour, soil brought in	1.52
e. Concrete blocks (ready made for use)	2.21 (for same wall thickness and area)

Source: Kateregga, 1982.

For the same wall thickness and area, the average unit cost of stabilised soil blocks is approximately 20 per cent to 70 per cent that of concrete blocks, depending on the organisation of production considered. The unit production costs for the stabilised soil blocks are estimated in the way described in the previous chapter. The estimates assume the use of a Brepak machine in a small-scale production process.

The production cost of stabilised soil blocks, and therefore their wholesale or retail prices should not be the only basis for comparison with other building materials. The house-owner or the builder is interested in the final cost of a wall, including the cost of building materials transported to the site, that of the mortar for the joints and that of labour. The latter two cost items are a function of the block size: the larger the size, the lower the labour and volume of mortar required for the same volume of walling. Thus, the labour cost and the cost of mortar for the building of a given volume of walling should be lower for concrete blocks than for stabilised soil blocks since the latter are usually smaller than the former blocks. Table VIII.3 provides the building cost of 1 m^2 wall (140 mm thickness) using, respectively, concrete blocks and stabilised soil blocks. The latter are produced under the four different organisations of production listed in table VIII.2. It can be seen from table VIII.3 that, in all cases, the total building cost per square metre of walling is lower for stabilised soil blocks than for concrete blocks (5 to 41 per cent lower, depending on the organisation of production considered).

Table VIII.3

Comparative costs of block walling

	Cost of 140 mm thick walling (with no surface finishing) - Kenya shillings per m^2 of wall area				
	Stabilised soil blocks [2] (290x140x100mm; 30.3 blocks/m^2)				Concrete blocks[1] (400x140x200 mm; 11.6 blocks/m^2) Ready made
Organisation of production (See table VIII.2)	a	b	c	d	e
Cost of blocks	13.64	22.42	37.27	46.06	64.15
2 per cent waste	0.27	0.45	0.75	0.92	1.28
Mortar for joint	9.64	9.64	9.64	9.64	4.83
Labour for laying	19.57	19.57	19.57	19.57	9.98
Total cost	43.12	52.08	67.22	76.19	80.24

[1] The cost of a concrete block is taken as 5.53 ksh.
[2] The unit costs of stabilised soil blocks are those shown in table VIII.2
Source: J.K. Kateregga, 1982.

It should be added that two other potential cost items could further increase the cost differential between concrete blocks and stabilised soil blocks in favour of the latter: the cost of rendering and that of the transport of concrete blocks to the construction site. In the case of good-quality stabilised soil blocks, rendering is often not necessary while it often cannot be avoided in the case of concrete blocks. Furthermore, concrete blocks are always produced at some distance from the construction site: transport costs must therefore be added to production costs. This is often not required for stabilised soil blocks. Thus, scarce foreign exchange may be saved whenever transport is minimised, since transport vehicles and fuel are often imported.

VI. CONCLUDING REMARKS

The preceding sections of this chapter have shown that, in general, the promotion of stabilised soil blocks in building construction should yield a large number of beneficial effects, especially in countries suffering from high unemployment and trade deficits. The promotion of good-quality blocks should also improve the standard of low-income housing and facilitate home ownership. This is the only building material which can be produced *in situ* if equipment and a limited amount of stabiliser can be made available. For example, housing authorities may organise the transport of a press and crushing/sieving equipment which could be operated by future home owners in self-help housing schemes. Training must, in this case, be provided by extension agents from housing development agencies. Alternatively, the equipment could be owned by a contractor, who would transport it from site to site in addition to other equipment, such as wheelbarrows and scaffolding.

In order to promote stabilised soil blocks and other forms of earth construction, the active involvement of housing authorities will be required in the following areas:

- revision of current building regulations to accommodate soil-based materials;

- inducements to future home owners to adopt stabilised soil blocks as the main building material; for example, the cost of the building permit may be reduced as an incentive;

- promotion of stabilised soil blocks through advertising, exhibits and pilot housing schemes;

- legislation recommending the use of stabilised soil blocks for some types of government buildings, schools, and so on;

- adjusting duties on imported building materials in order to make these less attractive vis-à-vis local materials;

- promotion, research and development in this field in order to maximise the use of local stabilisers and improve the quality of stabilised soil blocks; and

- organisation of training for the production of these blocks.

The implementation of the above measures should greatly contribute to making stabilised soil blocks preferrable to other building materials in terms of cost, availability and protection against adverse weather conditions.

APPENDICES

APPENDIX I

GLOSSARY OF TECHNICAL TERMS

Absorption	The taking up of water into a solid material. The quantity of water taken up by a stabilised soil block is a measure of the absorption of the block.
Adobe	Mud brick; hand-made, sun-dried, not fired;
Air void ratio	Ratio of volume of air voids to the total volume of the soil.
Alumina	Aluminium oxide (Al_2O_3)
Alveole	A pit or small depression as in the surface of a block.
Ambient	The surrounding natural environment, especially with reference to temperature, humidity and wind speed.
Arris	The edge where two external faces of a block meet.
Auger	Tool for boring a hole in the ground or for taking a sample of soil from the hole so made. It has a screw-like action when boring a hole.
Autoclave	The high pressure steam treatment given to some manufactured cementitious products to hasten the curing of the cement and to attain near maximum strength in a short time.
Bed face	The upper face of a block which is horizontal when laid in a wall.
Binder	The material which binds together separate particles; for example, cement binds sand in a mortar, also clay helps to bind together the coarser particles in a soil.
Bitumen	Natural mineral substance, normally black, melted by heat and dissolvable in organic solvents.
Bituminous emulsion	Dispersion of fine particles of bitumen in water.
Block	A rectilinear building unit which usually requires two hands or a special tool to lift it; in contrast to a brick, which may be lifted with one hand.
Bond	The laying of blocks in a regular pattern, in a wall, to obtain good strength and coherence in a wall. The vertical joints between the blocks in one course are not generally in line with those in the courses above and below.

Brick	A rectilinear unit made from clay, concrete, etc. with which walls may be built. Its size and weight are such that it can be lifted and laid with one hand. In plan view its length is usually twice its width (if thickness of one mortar joint is added to each direction) so that good bond may be obtained in walling.
Bulking of sand	A given weight of sand will vary in volume, depending upon the water content of the sand.
Calcine	Heat to elevated temperature. For example, limestone is heated to approximately 950°C to obtain quick lime.
Calcium silicate	A durable compound formed when lime reacts with silica, as for example during the autoclaving process in making calcium silicate bricks.
Clay	Natural minerals of many different types but consisting of very small particles, less than 0.002 mm. Because of their small size, when moist, they have cohesive properties and this permits deformation of a large mass into the desired shape (e.g. into the shape of a block).
Cleft	Split, such as may be found in a block.
Cob	Wall construction method in which moist clayey soil (and straw) are layed in lumps one upon another in courses without shuttering. The surface is trimmed flat as building proceeds.
Cohesion	The ability of a material to stick together; typically demonstrated by a moist clay.
Compaction	The packing together of particles of soil, under pressure, forming a more dense material.
Compaction pressure	The pressure applied, usually by a specially designed machine, to bring soil particles closer and reduce volume of air voids between them.
Compressive strength	The amount of compaction or squashing which a block can endure.
Course	A layer of blocks in a wall.
Crevice	A narrow crack.
Cure	Maintain environmental conditions so that a process may continue towards completion. Typically this would involve the maintenance of cement- or lime-containing materials such as stabilised soil blocks under moist conditions, so that the setting reactions of the cement may proceed or reactions may continue between lime and clay particles.

Dagga plaster	A mixture of clay soil, stabiliser and water used as an external rendering. It has medium resistance to rain.
Depreciation	Loss in value of equipment due to wear and tear over a period of time.
Dolomitic lime	Lime with approximately equal contents of calcium oxide and magnesia.
Down time	The amount of time during which equipment is not operating for a variety of reasons.
Dry compressive strength	The compressive strength of a block when it is tested in a dry condition.
Durability	The ability of a material to withstand conditions of service.
Fair-faced	Block walling of an acceptable standard of appearance and quality, without rendering or plastering.
Flash set	The premature setting of a cement, within a short period of time.
Formwork	Shuttering to contain soil, etc., during wall building.
Frog	Indentation formed during manufacture in one or sometimes both bed faces of a block.
Horse power	A measure of power, as, for example, the power of an electric motor. One hp is equivalent to 746 watts at unity power factor.
Hydraulic lime	A lime which sets under water.
Intermediate technology	A level of technology requiring neither unnecessarily high capital investment, sophistication and back-up services nor on the other hand, an unnecessarily high degree of manual labour. It generally utilises fairly simple processses and simple mechanical aids and it is largely synonymous with "appropriate technology".
Key	The roughness of a surface which helps a mortar to adhere to it.
Laterite	Highly weathered tropical soil, usually red in colour, sometimes containing hard nodules; rich in iron and aluminium oxides.
Lime	Two very different main forms of lime exist. Quicklime is calcium oxide, made by calcining limestone (or coral, or shells). Slaked lime is calcium hydroxide, made by careful addition of water to quicklime. They have similar uses in construction and in soil stabilisation but slaked lime is safer to handle and use.

Linear shrinkage	The decrease in length of a moist soil specimen as it becomes dry.
Lintel	A beam over a door or window, capable of supporting the wall above.
Liquid limit	The moisture content of a soil at which it ceases to be plastic and will just flow as a liquid.
Load bearing	A term applied either to the block or the wall built from blocks, indicating that it has to bear the load of all that is built above, without the benefit of a steel, concrete or wooden frame to take the load.
Loam	Sandy clay, often suitable for moulding into blocks, and having a low shrinkage.
Macro cracks	Large cracks.
Magnesia	Magnesium oxide (MgO).
Magnesian lime	A lime containing 5 to 40 (approx.) per cent of magnesia, derived from limestone containing magnesia.
Micro-cracks	Very fine cracks.
Mortar	A mix which may contain cement, lime, sand or soil, with water, for laying blocks in; it fills spaces between blocks and helps bond them together.
Mould	The metallic or wooden box in which soil is shaped into blocks. The action of shaping in a mould.
Moulding pressure	The pressure applied to the damp soil to force it into the mould, and compact the soil particles close together, to reduce the air voids ratio.
Optimum moisture content	The moisture content of a soil at which it can be compacted under pressure into the most dense block.
Ordinary Portland cement	A cement made by heating clay and limestone in a kiln at $1350°C$ (approximately) then grinding to powder the clinker which is formed. It is the material commonly termed "cement", although this is not a precise description of the material.
Pallet	A small board or platform, usually of wood, upon which one or more blocks may be carried.
Parallelepiped	Solid shape, contained by parallelograms.
Parallelogram	Four-sided rectilinear figure having its opposite sides parallel.
Permeability	A measure of how permeable a solid is.
Permeable	Allowing air or water to pass through, although solid itself.

Perpends	The visible vertical mortar joints between blocks in a wall.
Pisé de terre	Earth rammed between wooden or other formwork to make a wall in situ.
Plaster	The coating of cement/sand, lime, gypsum etc. applied to the block surfaces to give a smooth even finish to the wall. Sometimes, the term plaster refers exclusively to finishes indoors, especially when gypsum plaster is used.
Plasticity index	The difference between moisture contents at liquid limit and plastic limit.
Plastic limit	The moisture content of a soil at which it ceases to be plastic, and behaves more as a solid.
Pointing	A cement-based mortar trowelled into the raked-out joints between blocks after they have been laid.
Power factor	Efficiency of an electrical circuit.
Pozzolan	A natural or artificial inorganic material which in a wet mix at ambient temperature will react with lime, and set like cement. Pozzolans do not set by themselves.
Pulverised fuel ash	The fine particle-size ash remaining from burning of coal dust in some modern coal-fired electricity generating plants. It is a pozzolan, often referred to as PFA or flyash.
Punner	Heavy weight on bottom end of a pole, either for dropping on damp soil to compact it within formwork or for breaking up hard lumps of dry soil.
Rammed earth	Construction method for walls, in which earth is rammed down between formwork.
Release agent	A material applied to the surface of a mould to prevent the soil block sticking in the mould.
Render	Coating of durable cement/sand or other mix applied to wall surface.
Retard	Delay the time at which a cement or plaster starts to set.
Rice husk ash cement	A mixture of the ash of rice husks mixed with either lime or ordinary Portland cement.
Ring	The clear bell-like sound obtained when two well-compacted blocks are knocked against each other. Poorly compacted blocks produce a dull sound.
Rugosity	Roughness of a surface.

Sand	The smaller portion of the coarse material in a soil. Particle sizes from 2 down to 0.06 mm (British Standard definition).
Self-help	Producing materials such as blocks, or constructing buildings, using one's own labour. This may be by individuals or within small community groups.
Sesquioxides	The oxides of iron Fe_2O_3 and aluminium Al_2O_3.
Shrinkage	Reduction in size of moist soil as it dries.
Shuttering	Temporary structure, usually of wood, to retain soil as it is placed in situ to make a wall.
Silt	Particles of soil finer than sand, coarser than clay. Size 0.06 to 0.002 mm.
Skew	Not in a straight line.
Slake	Disintegrate by combination with water. In the case of a soil block, this will constitute the block's failure if it takes place. In the case of quicklime, it is a necessary process in making hydrated lime.
Solar gain	Heating up of an object by the radiated heat of the sun.
Spall	To split and splinter, pieces of a block thus becoming detached from the surface.
Specific density	The mass of a unit volume of material (measured for example in kg/m^3).
Stabilise	Improve properties of a soil by addition of other materials. Commonly, this improvement is obtained by making the soil more resistant to slaking and erosion by water.
Strain	Amount by which a body subjected to stress is deformed by that stress.
Strata	Layers of soil, sand, etc.
Stress	Amount of a force applied to a body.
Surkhi	Clay fired to temperature insufficient to develop full ceramic properties yet producing changes in the clay which increases its pozzolanicity. Surkhi is a traditional building material in India. The powdered material is used mixed with lime as a cement in mortar.
Swish	Mud walling. Swishcrete is mud with some cement added, for walling.

Temper	Leave clay soil in wet condition overnight, or longer, to enable moisture to permeate and improve plasticity.
Thermal mass	The property of a structure enabling it to store heat. Heavy building materials have greater thermal mass than lightweight ones.
Tonne	1000 kg.
Volume shrinkage	Decrease in volume of a moist soil specimen as it dries out.
Wattle and daub	A woven framework of branches and sticks which is smeared and interfilled with a wetted soil in order to form a wall.
Wet compressive strength	The compressive strength of a material immediately after it has been soaked in clean cold water for 24 hours.
Win	Obtain soil from the ground.

APPENDIX II

INFORMATION SOURCES ON STABILISED SOIL BLOCKS

ALGERIA

Département d'architecture,
Centre universitaire de Mostaganem,
B.P. 227,
MOSTAGANEM

AUSTRALIA

Division of Building Resarch,
CSIRO,
Graham Road,
Highett,
VICTORIA 3190

BELGIUM

CITADOBE,
Galerie Porte de Namur 5,
B.P. 79, Ixelles 1,
1050 BRUXELLES

Centre de recherches en architecture (CRA),
Université catholique,
Place du Levant, 1,
1348 LOUVAIN-LA-NEUVE

CRATerre BELGIUM,
57, Rue Franz Merjay,
1060 BRUXELLES

PGC - KULEUVEN,
Kasteel Arenberg,
3030 OUD HEVERLEE

BOTSWANA

Ministry of Local Government and Lands,
Private Bag 006,
GABORONE

Botswana Polytechnic,
Private Bag 0061,
GABORONE

The Botswana Technology Centre,
Private Bag 0082,
GABORONE

BRAZIL

University of Sao Paulo,
Butanta,
SAO PAULO

BURUNDI

Département de l'Habitat rural,
Ministère du Développement rural,
B.P. 2740,
BUJUMBURA

CANADA

Mc Gill University,
School of Architecture,
Minimum Cost Housing Group,
3480 University Street,
MONTREAL 101
(Quebec H3A 2A7)

CHINA

Architectural Society of China,
Baiwanzhuang,
BEIJING

COLOMBIA

Servicio de Intercambio Cientifico Documentación,
Centro Interamericano de Vivienda Planeamiento,
Apartado Aereo 6209,
BOGOTA D.E.

National Centre for Construction Studies,
Ciudad Universitaria C1145-CRA-30,
Edificio CINVA,
Apartado Aereo 34219,
BOGOTA

COTE D'IVOIRE

Fonds régionaux d'Aménagements ruraux,
B.P. 142,
06 ABIDJAN

DENMARK

Drostholm Products,
A/S-dk-2950 Vedbaek,
NR COPENHAGEN

DOMINICAN REPUBLIC

Centro de Tecnologia Apropriada para
la Vivienda Popular,
Apartado 20328,
SANTO DOMINGO

EGYPT

General Organisation for Housing, Building and Planning Research,
P.O. Box 1170,
El-Tahreer Street,
Dokky,
CAIRO

FEDERAL REPUBLIC OF GERMANY

German Appropriate Technology Exchange (GATE),
Dag Hammarskjöld-Weg 1,
6236 ESCHBORN 1

Forschungslabor fur Experimentelles Bauen,
University of Kassel,
Menzelstrasse 13,
3500 KASSEL

CRATerre,
Jahnstrasse 53,
6100 DARMSTADT

FRANCE

International Union of Testing and Research Laboratories (RILEM),
12, rue Brancion,
75737 PARIS CEDEX 15

Centre de Terre,
Lavalette,
31590 VERFEIL

CRATerre,
International Centre for the Research and
the Application of Earth Construction,
Centre Simone Signoret,
Quartier St. Bonnet Centre,
38090 VILLEFONTAINE

PISE, TERRE D'AVENIR,
7, rue Saint Pierre,
42600 MONTBRISON

GHANA

Building and Road Research Institute,
P.O. Box 40,
University,
KUMASI

Department of Housing and Planning Research,
Faculty of Architecture,
University of Science and Technology,
P.O. Box 40,
KUMASI

GUATEMALA

Building Information Centre,
Centro de Investigaciones de Ingeneria,
Ciudad Universitaria,
Zona 12,
GUATEMALA CITY

HUNGARY

Hungarian Institute for Building Sciences,
P.O. Box 71,
1502 BUDAPEST

INDIA

Central Building Research Institute (CBRI),
ROORKEE, Uttar Pradesh 277 672

Centre for the Application of Science and Technology to
Rural Areas (ASTRA),
Indian Institute of Science,
Mallesinaram,
BANGALORE 560012

National Building Organisation,
"G" Wing, Nirman Bhavan,
Maulana Azad Road,
NEW DELHI 110011

INDONESIA

Directorate of Building Research,
United Nations Regional Housing Centre,
84, Jajan Tamansari,
P.O Box 15,
BANDUNG

Building Information Centre,
20 Jalan Pattimura,
Kebayoran Baru,
JAKARTA SELATAN

IRAQ

National Centre for Construction Labs,
Tell Mohammed,
Mousa Bin Nesser Square,
BAGHDAD

ISRAEL

Building Research Station - TECHNION,
Israel Institute of Technology,
Technion City,
HAIFA

ITALY

CRATerre,
4, Via Roma,
33100 UDINE

JORDAN

Building Materials Research Centre,
Royal Scientific Society,
P.O. Box 6945,
AMMAN

KENYA

Housing Resarch and Development Unit (HDRU),
University of Nairobi,
P.O. Box 30197,
NAIROBI

United Nations Centre for Human Settlements - HABITAT,
P.O. Box 30030,
NAIROBI

LIBERIA

Soils/materials testing and research division,
Bureau of Technical Services,
Ministry of Public Works,
MONROVIA

MALAWI

Malawi Housing Corporation,
P.O. Box 414,
BLANTYRE

MEXICO

Information and Documentation Centre,
National Council for Science and Technology,
Insurgentes Sur 1677,
MEXICO 20 D.F.

NETHERLANDS

International Council for Building Research,
Studies and Documentation (CIB),
Weena 704,
P.O. Box 20704,
3001 JA ROTTERDAM

The Building Centre,
Bouwcentrum,
700 Weena, P.O. Box 299,
ROTTERDAM

PANAMA

Research Centre,
Faculty of Architecture,
University of Panama,
PANAMA

PAKISTAN

University of Engineering and Technology,
LAHORE

PERU

Instituto Nacional de Investigación,
Normalización de la Vivienda,
Pontificia Universidad Catolica del Peru,
Apartado 12534,
LIMA 21

CRATerre AMERICA LATINA,
Oficina de Coordinación Nacional e
Internacional,
Jr Ica 441-A, Of. 202,
LIMA 1

SENEGAL

ENDA-TM,
B.P. 3370,
DAKAR

SPAIN

CRATerre,
16 Rbla Luis Sampere 8,
CERVERA/LLEIDA

SUDAN

National Council for Research,
Housing and Engineering Unit,
P.O. Box 6094,
KHARTOUM

Building and Road Research Institute,
University of Khartoum,
P.O. Box 35,
KHARTOUM

SWEDEN

Swedish Association for
Development of Low-Cost Housing,
Arkitektur 1,
P.O. Box 725,
LUND 220 07

SWITZERLAND

Swiss Institute of Technology,
Institut HBT,
ETH Honggerberg,
8093 ZURICH

Institut universitaire d'Etudes
du Développement (IUED),
Rue Rothschild, 24,
1202 GENEVE

Swiss Centre for Appropriate Technology (SKAT),
Institute for Latin-American Research and
for Development Co-operation,
University of St. Gall,
ST. GALL

International Labour Office,
Technology and Employment Branch,
1211 GENEVA 22

UNITED REPUBLIC OF TANZANIA

Building Research Unit,
P.O. Box 1964
DAR-ES-SALAAM

CAMERTEC,
Centre for Agricultural Mechanisation
and Rural Technology,
P.O. Box 764,
ARUSHA

ARDHI,
P.O. Box 9132,
DAR-ES-SALAAM

THAILAND

Thailand Institute of Scientific and
Technological Research,
Building Research Division,
196 Phahonyothin Bangikhen,
BANGKOK 10900

Asian Institute of Technology,
Human Settlements Development Division,
G.P.O. Box 2754,
BANGKOK 10501

TOGO

Centre de la Construction et du Logement,
Cacavelli,
B.P. 1762,
LOME

TUNISIA

Ministère de l'Equipement et de l'Habitat,
Cité Jardin,
TUNIS

UGANDA

Building Research Unit,
Central Materials Laboratory,
Ministry of Housing and Public Buildings,
P.O. Box 7188,
KAMPALA

UNITED KINGDOM

Building Research Establishment (Overseas Division),
Bucknalls Lane,
Garston,
WATFORD WD2 7JR

Intermediate Technology Development Group,
Myson House,
Railway Terrace,
RUGBY CV21 3HT

Earthscan,
10 Percy street,
LONDON W1P 0DR

Intermediate Technology Workshop,
J.P.M. Parry and Associates Ltd.,
Overend Road,
CRADLEY HEATH, B64 7DD

British Standards Institution,
British Standards House,
2 Park Street,
LONDON W1A 2BS

Centre for Alternative Technology,
Llwyngwern Quarry,
Machynlleth,
POWYS, Wales

UNITED STATES

Volunteers in Technical Assistance (VITA),
1815 N.Lynn Street,
Suite 200,
ARLINGTON, Virginia 22209

Intertect,
P.O. Box 10502,
DALLAS, Texas 75207

Agricultural-Mechanical College of Texas,
Texas Transportation Institute,
A and M University,
COLLEGE STATION, Texas

Adobe Today,
P.O. Box 702,
LOS LUNAS, New Mexico 87031

International Foundation for
Earth Construction,
2501 M Street N.W.,
Suite 450,
WASHINGTON, DC 20037

WEST INDIES:
Jamaica

Building Research Institute,
34 Old Hope Road,
P.O. Box 505,
KINGSTON 5

Saint-Vincent
Christian Action for Development
in the Caribbean (CADEC),
P.O. Box 498,
KINGSTOWN

Trinidad and Tobago
University of the West Indies,
Department of Civil Engineering,
ST. AUGUSTINE

Caribbean Industrial Research Institute,
P.O. Box,
TUNAPUNA

ZAMBIA

Human Settlements of Zambia,
P.O. Box 50141,
LUSAKA

National Council for Scientific Research,
P.O. Box CH-158,
CHELSTON, LUSAKA

APPENDIX III

LIST OF EQUIPMENT SUPPLIERS AND MANUFACTURERS

Countries	Type of equipment
AUSTRALIA	
Australian Adobe Industries, Suite 4, Ormond House, 109 Yarra Street, GEELONG, Vic. 3220	Fully automated adobe Earth brick machine
BELGIUM	
Fernand Platbrood, 20, rue de la Rieze, B6404 CUL-DES-SARTS	Terstaram machine Clay crushing and sieving Mixers
CERATEC 228, rue du Touquet, 7792 PLOEGSTEERT	Cetaram machine
UNATA, Gud. Heuvelstraat 131, 2140 RAMSEL-HERSELT	UNATA machine (Modified CINVA-Ram)
J. Riffon, Rue J. Wilgot 6, 5220 ANDENNE	Pedal and lever operated press
BRAZIL	
Tecmor Equipamentos Mecanicos Ltda., Rua Visconde de Inhauma, 517 Sao Carlos, SAO PAULO	Tecmor machine Mixing Sieving
Torsa Maquinas e Equipamentos Ltda. SAO PAULO	Supertor machine
Industria e Comercio de Maquinas, Rua 3 de Dezembro, 33-50, Sala 55, SAO PAULO	CINVA-Ram machine
CAMEROON	
CENEEMA, B.P. 1040, YAOUNDE	CENEEMA machine (modified CINVA-Ram)
COLOMBIA	
Metalibec Ltda., Apartado Aereo 11798, BOGOTA	CINVA-Ram machine

Metalibec Ltda., Apartado Aereo 233, Na 1 157, <u>BUCARAMANGA</u>	CINVA-Ram machine
SENA Dirección general, Carerra 31, No. 14-20, Apartado Aereo 53329, <u>BOGOTA</u>	Maquina Machine

<u>COTE D'IVOIRE</u>

Abidjan Industrie, B.P. 343, 45, rue P. M. Curie, Zone 4C, <u>ABIDJAN</u>	ABI Block press (modified CINVA-Ram machine)

<u>DENMARK</u>

Drostholm Products A/S, 2950 Vedbaek, <u>NR COPENHAGEN</u>	Latorex plant and system equipment

<u>FEDERAL REPUBLIC OF GERMANY</u>

Ausbildungsverband Metall (AVM), Bernhard-Adelung Strasse 42, <u>6090 RUSSELSHEIM</u>	AVM block press (Modified CENEEMA machine)
Lescha Maschinenfabrik, Ulmer Strasse 249/251, <u>8900 AUGSBURG</u>	Lescha SBM press (Improved version of CLU 2000 machine)

<u>FRANCE</u>

SARET, B.P. 73 Route de Carpentras <u>84130 LE PONTET</u>	PPB Saret - Teroc machine Mixing, Sieving
CTBI, Zone industrielle, Rue du Grand Pré,<u>51140 MUIZON</u>	CTBI block press machine
RGF TERRE 2000, Système constructif, B.P. 113, <u>13160 CHATEAURENARD</u>	Terre 2000 hydraulic press Mixing
ALTECH, Rue des Cordeliers, <u>05200 EMBRUN</u>	Pact 500 block press
SOUEN Centre de Terre, Lavalette, <u>31590 VERFEIL</u>	TOB system - G.E.O. 500 semi-bloc (modified CINVA-Ram machine) (modified Winget machine)

RAFFIN,,   Dynaterre press
700, route de Grenoble,
B.P. 9 Domène,
38420 LE VERSOUD

GHANA

Department of Housing and Planning Research,   Tek-Block machine
Faculty of Agriculture,
University of Science and Technology,
KUMASI

GUATEMALA

Centro de Estudios Mesoamericanos sobre   CETA machine
tecnologia apropiada (CEMAT),   (modified CINVA-Ram to
Apartado Postal 1160   produce hollow blocks)
GUATEMALA CITY

Centro de Experimentación en   CETA machine
Tecnologia Apropriada,
15 Ave. 14-61, Zona 10,
GUATEMALA CITY

INDIA

ASTRA,   ASTRAM machine
Indian Institute of Technology,
BANGALORE 560012

Aeroweld Industries   ASTRAM machine
B9, Hal Industrial Estate,
BANGALORE

Joshi Industries,   Ellson-Blockmaster machine
Gayatri,,
Dr. Yagnik Road,
RAJKOT (Gujarat State)

Kathiavar Metal and Tin Workd Pvt, Ltd.,   Elsson-Blockmaster machine
9 Lati Plot,
RAJKOT (Gujarat State)

ITALY
Giza Spa.,   Plants for the production of
Sede Amministrativa,   stabilised earth blocks
42011 BAGNOLO IN PIANO (RE)

KENYA

Christian Industrial Training Centre (CITC), Clay crushing machine
Meru Road,
Pumwani,
P.O. Box 729935,
NAIROBI

Western College of Arts and Applied
Sciences (WECO),   WECO/CINVA-Ram machine
P.O. Box 190,
KAKAMEGA

Kenya Industrial Estates Ltd.,   Modified CINVA-RAM machine
Rural Industrial Development Centre,
P.O. Box 275,
MACHAKOS

SIHRA Engineering,   Bonner block making machine
Lunga-Lunga Road,
P.O. Box 16074,
NAIROBI

MEXICO

Estructuras desarmables, S.A.,   Blokorama press
Apartado Postal 1669,
MEXICO D.F.

NETHERLANDS

Centre for Appropriate Technology (CAT),   Prototype mechanised block press
Delft University of Technology,   (modified form of Winget machine)
P.O. Box 5048,
2600 GA DELFT

NEW ZEALAND

Frazer Engineering Co.,   CINVA-Ram machine
116 Tuam Street,
CHRISTCHURCH

PARAGUAY

CTA,   CTA block-press (modified
Facultad de Ciencias y Tecnologia,   CINVA-Ram machine to produce
Universidad Catolica,   3 blocks per cycle)
ASUNCION

PERU

CRATerre AMERICA LATINA,,   CRATerre Perou block press
JR Ica 441-A, Of. 202,   (modified Ellson machine)
LIMA 1   Handling equipment

SWITZERLAND

Robert Aebi, SA,   Automatic hydraulic press
8023 ZURICH

Bertrand S.A. Vevey,   CINVA-Ram press
24, rue de l'Union,
1800 VEVEY

Meili Engineering Co.,   Meili Mechanpress machine
Gewerbe Center Rothaus,
8635 DURNTEN

Maro Entreprise,   Maro block press (modified
Route de Suisse 95B,   CINVA-RAM machine)
1290 VERSOIX

CONSOLID A.G., Aechelistrasse 18, 9435 HEERBRUGG SG	CLU 3000 soil block plant
H.D. Sulzer, Institut für Hochbautechnik, ETH-Hönggerberg, 8093 ZURICH	Saturnia soil block press (range of modified CINVA-RAM machines)
Dieter Schmidheini, Weinbergstr. 29, 9436 BALGACH	Ecobrick 1000 (modified CINVA-RAM machine)

THAILAND

Southern Institute for Skill Development (SISD), Thai-German Project, P.O. Box 5, Kao Seng, SONGKHLA 90001	Soil block presser (modified CINVA-RAM machine)
Asian Institute of Technology, Human Settlements Development Division, G.P.O. Box 2754, BANGKOK 10501	Modified CINVA-Ram machine producing interlocking blocks

UNITED KINGDOM

Multiblock Ltd., Blackswarth Road, BRISTOL BS5 8AX	Brepak machine
Winget Limited., ROCHESTER ME2 4AA	Rotary table pressing machine
Zora International Company Ltd., 112 Power Road, LONDON W4 5PY	Zora machine
Intermediate Technology Workshops, J.P.M. Parry and Associates Ltd., Corngreaves Trading Estate, Overend Road, CRADLEY HEATH B64 7DD	Clay crushing sieving and handling equipment,
Baird and Tatlock Ltd., Freshwater Road, CHADWELL HEATH-ROMFORD	General laboratory equipment
Leonard Farnell and Co. Ltd., Station Road, North Mimms, HATFIELD AL9 7SR	Soil testing equipment and earth augers
G. Bopp and Co. Ltd., 102 Church Lane, EAST FINCHLEY	Sieve meshes

ELE International Ltd., Materials Testing Division, Eastman Way, HEMEL HEMPSTEAD HP2 7HB	General laboratory soil testing and compression testing equipment
Sutcliffe Speakman & Co., Ltd., LEIGH, (Lancashire)	Duplex Emperor mechanical brick making press
United Builders Merchants Ltd., Overseas Division, P.O. Box 78, Winterstoke Road, BRISTOL BS99 7EW	General builders merchants

UNITED STATES

Earth Technology Inc., 175 Drennen Road, ORLANDO, Florida 32806	Terrablock system
Bellow's Valvair International, 200 W Exchange Street, AKRON, Ohio 44309	CINVA-Ram machine
ULTRABLOC, P.O. Box 1363, CORRALES, New Mexico 87048	Ultrabloc impact hydraulic block press
Hans Sumpf Adobe Co., Fresno, California, Via:IFEC, 3282 Theresa Lane, LAFAYETTE, California 94549	Hans Sumpf Adobe block machine
Design Services, Box 2334, RUIDOSO, New Mexico 88345	Adobe master hand-operated adobe maker

APPENDIX IV

BIBLIOGRAPHY

Akroyd, T.N.W.: Laboratory testing in soil engineering (London, Soil Mechanics Ltd., 1962).

Bijlani, H.U.: "Rural housing in India", in Habitat International, Vol. 6, No. 4 (Oxford, Pergamon Press, 1982), p. 513.

British Standards Institution: Soils for civil engineering purposes, doc. BS1377 (London, BSI, 1975).

———: Methods of test for stabilised soil, doc. BS1924 (London, BSI, 1975).

———: Portland cement (ordinary and rapid hardening), doc. BS12 (London BSI, 1971).

———: Specification for masonry cements, doc. BS5224 (London, BSI, 1976).

———: Code of practice for structural use of masonry, Part 1 - unreinforced masonry, doc. 5628 (London, BSI, 1978).

———: Methods of test for chemical stoneware, doc. 784 (London, BSI, 1953).

———: Specification for precast concrete masonry units, doc. BS6073 (London, BSI, 1981).

———: Clay bricks and blocks, doc. 3921 (London, BSI, 1974).

Building Research Establishment: "External rendered finishes", in Digest 196 (Garston, BRE, 1976).

Butterworth, B. "The properties of clay building materials", Paper presented at a British Ceramics Society Symposium, Stoke-on-Trent, 1953.

Central Building Research Institute: "Cementitious binder from waste lime and rice husk", technical note no. 72, second edition (Roorkee, 1980).

Coad, J.R. "Lime-stabilised soil building blocks" in Building Research and Practice, Vol. 7, No. 2 (London & Paris, International Council for Building Research, 1979).

CRATerre (Doat, P; Hays, A; Houben, H) : Construire en terre (Paris, Editions Alternatives, 1983).

——— (Groupe Pisé): L'architecture de terre, région Rhône-Alpes (Bourg-en-Bresse, SME Résonances, 1983).

——— (Houben, H.; Guillaud, H.: Earth construction primer (Nairobi, United Nations Centre for Human Settlements, 1984).

——— (Hays, A.; Matuk, S.; Vitoux, F.: Seguir construyendo con tierra (Lima CRATerre, 1984).

———: "Traité de la construction en terre" dans Encyclopédie de la Construction en Terre (Marseille, Editions Parenthèses, 1987).

England R; Alnwick, D.: "What can low-income families afford for housing?" in Habitat International, Vol. 6, No. 4 (Oxford, Pergamon Press, 1982), p. 441.

Ferm, R.: Stabilised earth construction - An instructional manual (Washington D.C., Intrernational Foundation for Earth Construction, 1985).

Fitzmaurice, R.: Manual on stabilised soil construction for housing (New York, United Nations Technical Assistance Programme, 1958).

Grimshaw R.W.: The chemistry and physics of clays (London, Ernest Benn, 1971).

House, W.J.: Nairobi's informal sector, Working paper No. 347 (University of Nairobi, Institute for Development Studies, 1978).

Indian Standards Institution: Specification for lime-pozzolana mixture, doc. IS 4098 (New Delhi, ISI, 1967).

Kateregga, J.K.: A clinic built by the people for the people of Kabiro village in Kawangware-Nairobi (University of Nairobi, Housing Research and Development Unit, 1982).

———: Earth construction technology Kenya (University of Nairobi, Housing Research and Development Unit, 1983).

Kateregga J.K.; Webb, D.J.T.: Improved stabilised soil block for low-cost wall construction (University of Nairobi, Housing Research and Development Unit, 1985).

King. G.A.: "Gypsum products and their application in the Australian building industry", Paper presented at the New building materials and components symposium, Baghdad, 1979.

Lunt, M.G.: Stabilised soil blocks for building, doc. OBN 184 (Garston, Building Research Establishment, 1980).

Má R. L.: Manual of construction using the Ceta-Ram (Ciudad University Guatemala, Centre of Investigations of Engineering, 1981).

Okie, J.S.: Making soil blocks with a Tek-Block press (Kumasi, University of Science and Technology, 1975).

Ola, S.A.: The potentials of lime stabilisation of lateritic soils (Zaria, Ahmadu Bello University, 1977).

Parry, J.P.M.: Brickmaking in developing countries (Garston, Building Research Establishment, 1979).

Prescott, J.A.; Pendleton, R.L.: Laterite and lateritic soils, technical communication no. 47 (Farnham Royal, United Kingdom, Commonwealth Agricultural Bureau; 1966).

Sebestyen, G.: "Appropriate technology in the construction and building materials industry", Paper presented at the International Forum on Appropriate Industrial Technology, New Delhi, 1978.

Smith, R.G.: "Building with soil-cement bricks", in Building Research and Practice, Vol. 2, No. 2 (Paris, International Council for Building Research, 1979), pp. 80-89.

____: Small-scale brickmaking, Technical Memorandum No. 6 (Geneva, ILO, 1984).

Smith, R.G.: Rice husk ash cement (London, ITDG, 1983).

____: "Gypsum", Proceedings of a meeting on small-scale manufacture of cementitious materials (London, ITDG, 1974).

SKAT: Soil block making equipment (St. Gallen, SKAT, 1984).

Spence, R.: Making soil cement blocks (Lusaka, Commission for Technical Education, 1971).

____: Small-scale production of cementitious materials (London, ITDG, 1980).

____: Alternative cements in India (London, ITDG, 1976).

Stulz, R.: Appropriate building materials, Publication no. 12 (St. Gall, SKAT, 1981).

Swedish Association for Development of Low-Cost Housing (SADEL): Block making machines for soil blocks (Lund, SADEL, 1983).

UNIDO: The building materials industry in developing countries: An analytical appraisal, Sectoral studies No. 16, vol. 1 (Vienna, UNIDO, 1985).

United Nations: Report of Habitat, Conference on Human Settlements, Vancouver, 1976, doc. A/CONF.70/15 (New York, United Nations, 1976).

____: United Nations Global Review of Human Settlements, Conference on Human Settlements, Vabcouver, 1976, doc. A/CONF.70/A/1, (New York, United Nations, 1976).

Volunteers in Technical Assistance: Making building blocks with the CINVA-RAM block press (Mt. Rainier, Maryland, VITA, 1977).

Webb, D.J.T.: "Stabilised soil construction in Kenya", Paper presented at the International Conference on Economical Housing in Developing Countries, Paris, UNESCO, 1983, pp. 137-140.

West G; Dumbleton, M.J.: Wet sieving for the particle size distribution of soils (Crowthorne, United Kingdom, Road Research Laboratory, 1972).

World Bank: Focus on poverty (Washington, DC, 1983).

QUESTIONNAIRE

1. Full name..

2. Address..
   ....
   ....

3. Profession (check the appropriate case)

   Established stabilised soil blocks manufacturer....................../__/
   If yes, indicate scale of production................................

   Government official.../__/
   If yes, specify position..

   Employee of a financial institution................................./__/
   If yes, specify position..

   University staff member.../__/

   Staff member of a technology institution............................/__/
   If yes, indicate name of institution................................
   ....

   Staff member of a training institution.............................../__/
   If yes, specify...
   ....

   Other, specify..
   ....

4. From where did you get a copy of this memorandum?
   Specify if obtained free or bought..................................
   ....

5. Did the memorandum help you achieve the following:
   (Check the appropriate case)

   Learn about block manufacturing techniques you were not aware of /__/

   Obtain names of equipment suppliers............................. /__/

   Estimate unit production costs for various scales
   of production/technologies....................................... /__/

   Order equipment for local manufacture............................ /__/

   Improve your current production technique........................ /__/

   Cut down operating costs... /__/

   Improve the quality of blocks produced........................... /__/

   Decide which scale of production/technology to
   adopt for a new stabilised soil block plant...................... /__/

   If a Government employee, to formulate new measures
   and policies for the construction industry....................... /__/

   If an employee of a financial institution, to assess
   loan requests for the establishment of a
   stabilised soil block plant...................................... /__/

   If a trainer in a training institution, to use the
   memorandum as supplementary training material.................... /__/

   If an international expert, to better advise counter-
   parts on stabilised soil blocks manufacturing technologies....... /__/

6. Is the memorandum detailed enough in terms of: Yes   No

   - Description of technical aspects.........................____.....____

   - Names of equipment suppliers.............................____.....____

- Costing information..____.....____

- Information on socio-economic impact.....................____.....____

- Bibliographical information..............................____.....____

If some of the answers are 'No', please indicate why below or on a separate sheet:

....
....
....

7. How may this memorandum be improved if a second edition is to be published?..
....
....

8. Please send this questionnaire, duly completed to:

   Technology and Employment Branch
   International Labour Office
   CH-1211 <u>GENEVA 22</u> (Switzerland)

9. If you need additional information on some of the subjects covered by this memorandum, the ILO will do its best to provide the information requested.

Other ILO publications

Technology Series

The object of the technical memoranda in this series is to help to disseminate, among small-scale producers, extension officers and project evaluators, information on small-scale processing technologies that are appropriate to the socio-economic conditions of developing countries. ISSN 0252-2004

Tanning of hides and skins (Technology Series – Technical Memorandum No. 1)

Provides technical and economic information concerning the tanning of hides and skins at scales ranging from two hides per day (a typical rural tannery) to 200 hides per day. Six alternative tanning technologies are described, from a fully mechanised 200 hides per day project to a highly labour-intensive two hides per day project. Subprocesses are described in great detail, with diagrams of pieces of equipment which may be manufactured locally. A list of equipment suppliers is also provided for those pieces of equipment which may need to be imported.

The memorandum on the tanning of hides and skins is, to some extent, complementary to that on the small-scale manufacture of footwear. ISBN 92-2-102904-2

Small-scale manufacture of footwear (Technology Series – Technical Memorandum No. 2)

Covers the small-scale production of footwear (shoes and sandals) of differing types and quality, providing detailed technical and economic information covering four scales of production ranging from eight pairs per day to 1,000 pairs per day. A number of alternative technologies are described, including both equipment-intensive and labour-intensive production methods. Subprocesses are described in great detail, with diagrams of pieces of equipment which may be manufactured locally. A list of equipment suppliers is also provided for those pieces of equipment which may need to be imported. ISBN 92-2-103079-2

Small-scale processing of fish (Technology Series – Technical Memorandum No. 3)

This technical memorandum covers, in detail, technologies that are suitable for the small-scale processing of fish: that is, drying, salting, smoking, boiling and fermenting. Thermal processing is described only briefly, as it is used mainly in large-scale production. Enough information is given about the technologies to meet most of the needs of small-scale processors. Two chapters of interest to public planners compare, from a socio-economic viewpoint, the various technologies described in the memorandum and analyse their impact on the environment. ISBN 92-2-103205-1

Small-scale weaving (Technology Series – Technical Memorandum No. 4)

Describes alternative weaving technologies for eight types of cloth (four plains and four twills) of particular interest for low-income groups in terms of both price and durability. The monograph provides information on available equipment (e.g. looms, pirning equipment, warping equipment), including equipment productivity, quality of output, required quality of materials inputs, and so on. A methodological framework for the evaluation of alternative weaving technologies at three scales of production is provided for the benefit of the textile producer who wishes to identify the technology/scale of production best suited to his own circumstances. A chapter of interest to public planners compares, from a socio-economic viewpoint, the various weaving technologies described in the memorandum. ISBN 92-2-103419-4

Small-scale oil extraction from groundnuts and copra (Technology Series – Technical Memorandum No. 5)

Covers, in detail, various technologies for the extraction of oil from groundnuts and copra: baby expeller mills, small package expellers and power ghanis. Three main stages of processing are considered, namely the pre-processing stages (drying, crushing, scorching), the oil extraction stage and the post-treatment stages (filtering, cake breaking, packaging, bagging). The economic and technical details provided on the stages of processing should help would-be or practising small-scale producers to identify and apply the oil extraction technology best suited to local conditions. A chapter of interest to public planners compares small-scale plants and large-scale plants from a socio-economic viewpoint and suggests various policy measures for the promotion of the right mix of oil extraction techniques.

ISBN 92-2-103503-4

Small-scale brickmaking (Technology Series – Technical Memorandum No. 6)

Provides detailed technical information on different brickmaking techniques and covers all processing stages, including quarrying, clay preparation, moulding, drying, firing and the testing of produced bricks. The techniques described are mostly of interest to small-scale producers in both rural and urban areas. The processes and equipment are described in great detail, including drawings of equipment and tools which may be produced locally, floor plans, labour and skill requirements, materials and fuel inputs per unit of output. A list of equipment suppliers from both developing and developed countries is also supplied with a view to assisting the would-be brickmaker to import the required equipment. A chapter of interest to public planners compares, from a socio-economic point of view, the various brickmaking techniques described in this memorandum. ISBN 92-2-103567-0

Small-scale maize milling (Technology Series – Technical Memorandum No. 7)

Provides detailed information on alternative techniques for the production of whole meal, bolted meal and super-sifted meal. It covers all processing stages, including grain preparation, shelling, milling, sifting and packaging. The techniques described are mostly of interest to custom mills and small-scale merchant mills located in rural and urban areas. The first chapter of the memorandum provides in-depth information on the various effects of different maize-milling techniques, including the nutritional values of various kinds of maize meal, employment generation, foreign exchange savings, and so on. It also provides some general guide-lines for the formulation and implementation of measures for the promotion of appropriate maize-milling techniques. ISBN 92-2-103640-5

www.ingramcontent.com/pod-product-compliance
Ingram Content Group UK Ltd.
Pitfield, Milton Keynes, MK11 3LW, UK
UKHW051524180426
11947UKWH00018B/1567